down to earth

down to earth

the life and views of

TED MOULT

GOLDEN EAGLE - LONDON
29 OLD BOND STREET

Published by
GOLDEN EAGLE PRESS

Editorial and production office
55 VAUGHAN WAY
LEICESTER
LE1 4NT

ISBN O 901482 18 8

Printed in England for Golden Eagle Press Ltd., London

DEDICATED TO ALL
THOSE CREATURES ON
WHOM I HAVE LEANED

Contents

Illustrations

Foreword

ONE day, sometime in 1972, I had a call from the B.B.C. Would I take part in the next series of the radio panel game "Treble Chance", captaining one set of teams against another set led by Ted Moult?

I accepted without hesitation. I had heard several editions of "Treble Chance", and I had seen and heard Ted Moult often enough; but I had never actually met him. When our paths crossed (as happened very consistently for the next few months, and as will, I hope and believe, continue to happen in the future) I found him exactly as I had expected. There are some people, often seen on television, who prove to be immensely disappointing when met face to face. Ted Moult did not come into this category at all. Moreover, he is a man of many talents; and this is why his book is so interesting.

On Page 1, Ted says "I know I'm ordinary". This is what he sincerely believes. In fact, he isn't ordinary at all. He is, first and foremost, a farmer; and there can be few people in England today who know more about the practical side of farming. When discussing, say, the peculiar habits of cows, the pros and cons of synthetic food, or the best way to cure a mysterious ailment in pigs, he is writing as an acknowledged expert. But there are many people—including myself, I must admit—who are not farm-minded by inclination; and if Ted had written "just another book about the farm" it would have aroused little comment. What he has, in fact done, is to produce a splendid series of essays, and his topics range from Darwinism to drugs, the situation in Hong Kong, pollution, sex education and the Ashbourne Football Match. (The latter is something that I really must see before I die. I doubt whether I will understand it, despite Ted's masterly description; but then, as a mere cricketer, I don't understand Rugger or Soccer either).

There are folk today who set themselves up, quite consciously, as "instant experts". They do some selective reading, and then ask to be accepted as authorities in their own right. Nothing could be further from Ted Moult's mind; but quite apart from his versatility and his wide general knowledge, he has something which is so often lacking in the world today—common-sense. To all his subjects he brings a wholesome, sane outlook, so that the title of the book, **Down to Earth,** is by no means inappropriate. Whether or not one agrees with all he says, nobody can question the reasoning behind his views. I would be the last to claim that I hold views which are always identical with his; for example, I am much more definite about fox-hunting, for which I can see no conceivable excuse in what is supposed to be a civilised community ! But whatever Ted Moult says is worth hearing, and whatever he writes is worth reading. As you go through this book, I am quite sure that you will agree with me.

Finally, he is never pretentious; and I believe that the sign outside his farm in Derbyshire symbolizes his outlook. Ted sells potatoes. Many people would put up a notice saying, for instance: EXCELLENT POTATOES: FIRST-CLASS QUALITY: REASONABLE PRICES: FULL SATISFACTION GUARANTEED. Not Ted. His sign reads, simply: GOOD SPUDS.

As a broadcaster and speaker, Ted Moult has given many people much pleasure—and much food for thought. I am therefore particularly glad to be writing the Foreword to this book—the first from his pen, but certainly not his last.

PATRICK MOORE

Selsey, August 29th, 1973.

1

Calling a Spade a Spade

THE first question that anyone should ask himself, when starting to write a book, is what right or qualification has he to inflict such an opus on the poor, unsuspecting public? What right have I, as an ordinary bloke, to express opinions on a variety of subjects ranging from A for Agriculture to Z for Zany behaviour; and setting aside false modesty, who will want to read it anyway?

If one argued the pros and cons objectively, the first conclusions might well be that politics are for the politicians, statistics for the mathematicians and views on sex and birth control the sole province of the pop-doctors of the avante-garde. I suppose that thoughts along these lines have stopped me in the past from committing pen to paper. Then why now? Do my views carry more conviction, and are a greater number of people interested in them, than would have been some years ago? If I am what the majority of the media call me, an ordinary sort of a chap, I feel the answer is "yes".

I know I'm ordinary. Once as I was on my way to take part in an *Any Questions* programme, one of those chaps with big gloves, who couple together the carriages on Waterloo Station remarked "I like to see you on the telly, Ted – You're so *ordinary*".

The British should be thankful that we live in a country which allows free speech and uncensored information, sharing this privilege with less than a quarter of the globe's independent sovereign states. Moreover, it is the mark of a mature nation when a large percentage of its subjects are articulate, and opinions of all shades are freely available, although it wasn't until the train passed through Andover Junction that I had worked out the railwayman's remarks.

For too long though, we have been subjected to a mass of so-called expert opinion, much of it coming from people who appear to live in cloud cuckoo land. I also criticise the instant opinions, too often volunteered by the most inexpert characters. Surely a greater sense of discernment should be exercised. It's time we got down-to-earth, and who better to dig the dirt than me?

There is a lot of truth in the old adage that you're nearer God's heart in a garden than anywhere else on earth. A sense of oneness with nature is something that a farmer, or those who work close to the soil in all seasons can fully understand. I don't want to litter this book with proverbs, but it's worth remembering that words which have stood the test of time contain more than a grain of truth. As, for example, the one about the spectator seeing more of the game than the players. It is when one is away from it all that much falls into perspective. I have seen a great deal of the game both as a player and a spectator; the glitter and fame on the one hand, and the farm muck on the other (sometimes muck on both hands!)

Such a sharp contrast, often occurring abruptly, and always intermittently, has been the experience of few men. My field of vision has flashed from the fields and open sky to television networks and back again, from *What's My Line* to the dairy herd. I've often driven back from a T.V. programme late at night to find a note pinned to the dairy door "Looked at Deirdre 10.30 p.m. No calf yet. Milk Fever might be coming on." So on with the overalls and wellies, armed with a bottle of calcium solution to stick under Deirdre's skin or into her jugular vein, if I can find it!

If you feed two bullocks on the same diet, they cannot be relied upon to respond in exactly the same way, although most of them do. The same philosophy can be applied to people. It is alleged that some who have television fame thrust upon them become changed personalities, are totally carried away by this adventitious happening, and lose the common touch.

People have often asked me if this was so, and how T.V. had affected *me*. "Ah", said I, with practised timing, "I don't know how affected I was to start with." As for losing the common touch, whatever that is, I've never claimed to be anything but myself.

Personally, I believe television to be as good a mirror of human behaviour as one is likely to see. Any idiosyncrasies are accurately reflected. On meeting television people face to face, I have not found that they varied from the impressions I had already formed of their character, after seeing them previously on the box.

Just in case you've never heard of me, or even if you have, I'll give you a quick run-down of my history, up to the time a sheer fluke got me on the air. You may not find my beginnings very exciting, but here they are for the record. This will arm you with verbal dung to fling at me when you disagree with my views. Here goes!

I was born on the 11th February, 1926. Those of a mathematical turn of mind can work out how old I am. My father was a draper, my mother a very independent woman, whose only gift to the human race was me! Before her marriage she claimed the distinction of knowing the telephone numbers of everyone on the local subscribers' list by heart. Mind you, being in charge of the Derby Telephone Exchange gave her a head start. Her maiden name was Elsie May Walker, hence the reason for mine being Edward *Walker* Moult.

When I was old enough to be educated, I was sent to a Day Grammar School – the oldest in the town, dating back to 1160. I remember still that the school fees were six pounds a term. Forty per cent of the pupils were scholarship boys. I was not, as Pa must have had enough spare cash for the fees.

Being an only child, I suppose that I must have been spoiled, although I didn't notice it at the time. Maybe it was a subconcious recognition of this theory that resulted in my producing my own large family, and made numbers easy to reconcile. When the war started in 1939, evacuation was the call of the day as far as the young were concerned, and I, and a whole gaggle of my school mates were sent off to the Peak District. It was a lovely summer, and the war seemed far away from the blue skies above Overton Hall. This Youth Hostel became our home. I was thirteen at the time, and it all seemed very much a game, yet a grim game it turned out to be for many young men only a few years older than myself. Many of them found the skies anything but bright.

Impressions stick easily in one's youth. Maybe the age at which the blotting paper sucks up outer effects most easily is under ten, or even seven, but in the teens it is possible to actually remember being conditioned by one's environment.

At first, all we boys lived in the Hall, and lessons were held in the oak panelled splendour of former dining rooms, libraries and corridors. The overflow was accommodated at the Parish Room in Ashover and on fine days in the garden. Living, rationing, laundry and ablutions needed a lot of attention and organisation.

In our spare time we held parades in what was left of the Officer's Training Corps, later renamed the Junior Training Corps. We frequently went on patrol, up and down the Derbyshire hillsides, thus emulating the main activity of the British Army in those first few months of the phoney war. We furiously dug patches in the old neglected hall garden, hoping that the acid soil would produce a self-sufficiency in vegetables, and were extremely conscious of shortages. I fell down the stairs one day, whilst carrying fifty hard-boiled eggs. They weren't meant to be hard boiled, but this culinary error was all to the good. They were still edible, and even ready cracked!

Looking back at any experience, one tends to remember the most pleasant episodes, and while I'm not pretending that our existence was particularly spartan, it was rigorous on occasions. However, no boys were lost – except Eric Holmes, who on one of these patrols gave a frantic shout and disappeared into the dark bowels of the earth!

While traversing a difficult scree, he had lost his hand hold and slithered down a slope into a disused leadmine. The body, still mercifully articulate, was stuck on a ledge about sixty feet below the surface. We borrowed a wagon rope from the nearest farm and were able to haul him out without further disaster, but Eric's elbows and knees were painfully raw for a long time.

It wasn't long before some of the boys were billeted out, and I became a working guest of a Mr. Mycock of Ravens Nest Farm, Milltown, Ashdown, near Chesterfield. The memories flood back of those days, richened possibly, by nostalgia. In any case, I was there until 1941 and got my first real experience of the good earth.

We lived in a converted loft, vacated by some hens, and this served as a dormitory. It was a very cold winter! We had a latrine bucket, painted white round the top, for nocturnal use. One night some wag painted a white circle on the floorboards and removed the bucket. Fun in the blackout no doubt!

In the spring, the school moved to Amber Valley Camp, and *I* fell victim to a kind of practical joke. Spotty chicken pox, to which all kids should be exposed before the weaning age, grabbed me, and this disease, plus an ear infection, kept me out of the classroom for about four months.

Despite the scholastic setback, due to ill-health, I successfully took my School Certificate in 1941. It was in the same year, during the summer holidays, that at the age of fifteen, I got a job on a farm near Loughborough, Leicestershire.

In case you're getting restless with this mundane history and data, I'm going to throw in a few highlights at random. The incidents themselves are of little importance, but they are the things that spring to my mind, and perhaps therefore worthy in the telling of the tale.

One incident that still makes me smile, concerns a fellow named Bill Ellis who was also at Overton Hall. He was one of the few, like myself, who took a real interest in farming. In a spate of enthusiasm, he wrote to a milking machine company, asking for free pamphlets, as so many boys do, hoping that this literature would add to his 'expertise'. It was an early attempt at oneupmanship, but rebounded with a vengeance. One day, when the class was half way through a French lesson, a representative from the milking machine company arrived. He didn't come alone! Oh, no! He brought the entire new fangled equipment with him, hoping for a sale. There were red faces all round, not least of all that of young Bill Ellis. Ah, well, Bill, the incident has become part of a book after all these years, so perhaps it was worth it.

My interest in farming subjects continued to expand, and it was whilst I was at Amber Valley Camp, that I got a credit in mathematics. Figures have always interested me. In addition, history and languages were other pet subjects, but with an eye on a farming future, I decided to change to science, botany, biology, chemistry and the like, whilst still continuing an all-round education.

By the end of term, I decided I'd had enough of formal schooling and left at Christmas. Early the following year I went to work on a farm at Loughborough until I was seventeen. This episode in my life was brought to an abrupt halt, and an unpleasant one too. Lifting a huge bag of spuds into a cart caused a hernia and ten days in Ward 10. During convalescence I took a course for a Bachelor of Science degree, and when I was able, I studied part-time at the local Technical College. Then I got bored, and abandoned the idea. It isn't just today's youth who get fed-up. I think everyone in their teens goes through a stage of uncertainty about the future. It's a mixture of dreams, aspirations and dashed hopes. This should be borne in mind when looking at today's scene. It hasn't changed all that much.

Before we leave the Loughborough saga, there are two other items worth telling. The first is that during the time I was on the farm in that area, my grandfather came to lend me a hand. He was a character with a capital 'C', the sort that are few and far between today. This man had a big influence on my life, and still does in spirit, right up the present minute.

His name was Edward Bailey Walker, and like me, he was known of course, as Ted. He was a retired cattle dealer and meat buyer. At the time he came to give me a willing hand, he was living with my parents. Hours of my childhood had been spent with him in cattle lorry cabs. He had a sharp eye for a fat beast, and a quick, ready-reckoner like brain. He once bought fifty bullocks in one lot; the vendors took out the best beast and substituted another which grandfather had not then seen. Watching them arrive, the old man knew the switch had been made, even out of all those animals. Whether he missed the absentee or spotted the newcomer I'll never know.

More than being good at his job, he was a master of droll philosophy, and the down-to-earth approach. He came from Rutland, the smallest county in England and was very proud of it. That same pride still lives on in the county as witness the successful fight made by this brave little area to retain its geographical independence. Rutland is known far and wide for its three horseshoes crest, but that's another tale.

My grandfather was what was known in those days as 'very straight'. This may be best understood by telling a simple story concerning him. In his early days he was a butcher's boy at a shop in Derby. The man in charge was very suspicious of human nature, and to test the honesty of his young assistant he purposely left half-a-crown in the washing up water. Ted found it and returned it at once. The sly trick so disgusted a visiting butcher, who had witnessed the whole affair, that he displayed his objections in a most physical way, by taking the coin and cleaving it in twain with a meat axe.

To get back to the thread of the story, I saw a job advertised in the Farmer and Stockbreeder, breaking in horses. I took it, and it turned out to be in Quarnden, a village near Derby. There I lived with three Irishmen, three Italians and four daughters. I speak figuratively, of course. They were part of the farming community.

One thing that hadn't arrived in those days was the permissive society.

After two years the bug of dissatisfaction hit me again, and I took another farming job at a mental hospital at Mickleover where my parents then lived. There I worked until I was twenty-two years old.

In 1949 my ambitious urge led me into an attempt to set up on my own. They say that fools rush in where angels fear to tread, but this is one corny old saying that fortunately didn't work itself out. In fact, from that date to this I've been my own master.

Poplars Farm, Sinfin, became the home of mother, father and myself, and it was now up to me to prove my theories in practice. To do this meant working a sight harder than I did when employed by other people.

The little surplus energy that I had left over from heavy working days went into Rugby football. Whether the game really suited me, or whether it provided an outlet for my new-found frustrations and temper, I shall never know. From 1950 to 1954 I played for Derby Rugby Club (Third XV, unless some violent epidemic gave me a chance higher up) in the front row of the pack, and enjoyed every minute of it. *We* called it Rugby. The phrase 'coarse Rugby' had not then been coined, otherwise I would have known how to more accurately describe the game I was supposed to be playing.

I got married on November 27th, 1954. It was a very rough night – the Goodwin lightship blew over.

These days, young kids marry far too early, in my opinion. I sometimes tell my own lot that smart people do not tie themselves down too soon. Perhaps if more late marriages took place, the soaring birth rate might be steadied down – who can say?

I was twenty-eight when I married Marie. She was seven years younger, which is in accordance with an old belief that a desirable union is made if the woman's age is half the age of the man, with seven years added on. Marie was a staff nurse at the local Infirmary working in the operating theatre. She left the hot atmosphere of surgery for the cold comfort farm, which immediately took on a new air, as the old house felt the touch of feminine enthusiasm.

Looking back, it seems strange that we didn't even consider that she should carry on with her chosen career. Fortunately, it is now outdated to assume that every young woman should spend all her energies on domestic work and rearing kids. Surely this is a matter of individual choice.

Since then our family has grown by the addition of six offspring – three boys and three girls.

William was the first, born in August 1955, next came Anna in January '57, followed by Tommy in May '58, and Patrick in March '61. Next in line was Amelia who was born in July '62 and Ellen-Mary the latest addition, arrived in April, 1968.

Next we come to an episode that shows how big a part chance plays in our lives. A pure accident of fate made the difference between my becoming known, or remaining an obscure farmer for the rest of my life.

As my grandfather used to say, the ingredients for success are intelligence, initiative and influence, the theory being that if you possess any two of these factors you can get by, but with all three you should go into orbit! I would not attempt to analyse myself, unlike Freud (Sigmund not Clement), but would suggest that another necessity is being in the right place at the right time. Luck, if you like. This paid off for me in 1958, when I accidentally became a television performer, through a funny application to go on the radio. One of the results has been that I have been able to see far more of the Earth's crust than ever I would have imagined, when in 1954, I was in Paris on a week's honeymoon.

As well as all the E.E.C. countries, I have been fortunate enough to visit more exotic spots like Muscat and Oman, Chicago, Malacca and Macao, to name but a few.

The old saying that travel broadens the mind is certainly true, even though there is a general uniformity in airports, hotels and the like, all over the globe. I have always been a bit nutty about maps and charts, and my own piece of luck transformed the pages of the gazeteer into real earth, water, and of course, human beings.

2

A Job or a Way of Life?

WHEN I look back to 1949, the year I became my own boss, I'm immediately aware of what was possible then and impossible today. To have considered becoming on entrepeneur on £1,200, my total worldly wealth, might, one would think, excite my vanity.

I wouldn't have possessed that much cash had not an old aunt, widow of a Norfolk gamekeeper who, in his life, never earned more than £104 per annum and a new suit, left me a few hundreds in her will.

I knew that I wanted to farm and up came an opportunity when the fifty four acre Poplars Farm came up for sale.

Evidently my youthful zest impressed my father who sold his house and obtained a mortgage on the farm. The whole family moved to the new house, me being tenant and Pa the landlord.

Taking into account the disease of the 1970s – inflation – there was not much cash to spare for anything but bare essentials. I wanted to keep cows, but more to the point, I needed cows to keep me, so I used the 'ready' to fill up the shed with milkers, and purchased the minimum of equipment, mostly of the second hand or vintage variety, from farm sales or friendly neighbours. I bought two horses for £120 the pair, a job lot of saddles and bridles, and two shop soiled, tinned milkpails and milking stools.

My all time low investment was for a single furrow plough with wooden handles (or stilts) for four shillings which I actually used to break up a four acre worn out pasture. This was in preparation for the sower, which was me, going forth to sow oats by hand from a hopper slung round my neck.

I wish I still had that plough, because what was obsolete then is antique today.

It took a lot of handling, but the real struggle was between the two horses George and Harold, renamed by me, of course, from original drab titles like Boxer or Prince.

George always wanted to be in 'top gear' while Harold preferred to be in bottom, consequently, to even up matters I had to tie George back by means of a rope to Harold's collar.

Before long, I realised that some sort of mechanisation would make life much easier, and avarice reared its head when I worked out I could keep two more cows if I got rid of the horses. So, after twelve months I said 'Good-bye' to George and Harold, and bought my first tractor, which was a petrol-driven Ferguson.

Also, I purchased a milking machine, not only for milk production, but it had occurred to me that the cows might be happier. The decision was made after I had returned to the farm after playing Rugby one Saturday afternoon, and had fallen asleep on the milking stool under a cow. I was awakened by the cow, who apparently thought the operation so tedious that she kicked me out! Actually, I was overworking without realising it. I had had no real holiday for years.

Referring to the tractor, it might be worth reminding the reader that the great Irishman, Harry Ferguson, was responsible for one of the most significant steps in farming history, by his development of the tractor with hydraulic-mounted implements.

Today's children are probably so used to this method, through playing with their farmyard toys, that the old idea of using a horse instead of a tractor to cultivate the land seems quite abnormal.

With so much public attention today focussed on the world's dwindling energy resources, I wonder whether in the next century we shall be back to draught animals, unless nuclear power or something equally as revolutionary is made available to us.

Of all the jobs on a farm, ploughing gives me the most satisfaction. There is something poetic about turning the pale stubble into rich brown earth, seeing the seagulls following the plough and leaning back on their heels to tug at the worms exposed by the fresh furrows.

Yet, this aesthetic pleasure is increased with horses. For one thing the ploughman is looking to the front to watch the operation in contrast to the tractor driver who must look over his shoulder. Secondly, without any engine noise, one can hear the stones and soil scraping over the breast of the plough. And, what is better, the smell of fresh earth or the doubtful perfumes of petrol or diesel oil?

"Wee sleekit cow'ring timerous beastie,
What maks thee rin awa sae hastie."

As one inadvertently evicts a field mouse and her family, and sees the little creatures scurrying up the furrows, one realises what a true poet of the people was Robert Burns.

I became fairly proficient at horse ploughing when I was working for William Naylor at Vicarwood Farm as a youth, when one of the objects of working the horses as much as possible, was to break them in as draught animals for the Railway. Sometimes we would have them three in a line, the most inexperienced in the middle, and sometimes three abreast, the centre one walking in the furrow. On occasions we worked them four abreast, which necessitated seven swingle trees (wooden stretchers between the chains), between the animals and the harrows, or the plough. If a young horse in the middle got his leg over the chain it was certainly no dead man's job to lift it back.

Once, and only once, did I enter a Ploughing Match – in the Boys' Class at a local event. I managed to win second prize which pleased me no end, although there were only four lads in the contest. My ploughing was only average, but I could, and still can, for that matter, draw a straight rig, driving my horses at a line of peeled sticks set up along the field. Match contestants had to set up one rig where they started, and finish with one furrow.

Livestock at Poplars Farm consisted of a herd of milking cows, poultry and pigs. The pigs consisted of twelve sows and one boar, which we called 'Henry the Eighth'. They were mainly Large Black or Saddleback pigs, the latter breed being very popular where I was concerned because of their large ears. Their ears flopped over their eyes, and as they couldn't see where they were going, they were easily contained in the fields, particularly by electric fences. Apart from this admirable trait, the two breeds, when crossed with a Large White or Landrace boar, produced a litter of 'Blue and White', which, although not now quite so popular, fetched a top price in the weaner market in those days.

The criterion with pig breeding is making sure firstly that enough pigs are born, and secondly that as many as possible live. I can't say to maturity, because pigs usually die happy before they reach that stage in development, but at least to an age when they can be sold.

Nowadays, on large pig breeding units, the sows usually farrow in batches and are put in crates to make sure that they cannot crush the little piglets while they are very small. When I started farming, it was quite an innovation to use an infra-red lamp suspended over a fenced off area of the farrowing shed. Here, the baby pigs could take refuge and also maintain their body temperature, to prevent a kind of frozen paralysis.

If the piglets kept warm by the mother's side, they were very vulnerable to any lurch or movement on the mother's part, probably due to the pain of giving birth, and it's no joke for a newly born pig to have a hundredweight or so of Saddleback descending upon it. After the bulk of the litter was born, however, it was quite significant that the sow would make low grunts as if she was talking to her children, and from then on there was very little trouble.

I was reasonably lucky with my pigs. Sometimes there were a small number of very strong pigs, and other times there would be a very large litter to offset the inevitable losses. I remember the night before I married, one of my sows produced twenty pigs of which six were eventually reared. I still wonder whether it was significant that this massive effort in fecundity occurred on my wedding eve.

The most disappointing experience was a sow that produced six pigs without any milk to feed them. Those who think feeding babies from a bottle is difficult should try pigs. The little swine are the very devil to feed with a bottle. Not only do they appear determined to die of starvation, they have teeth like needles, and it is very awkward to get the teat in their mouths. Unlike babies, they have sharp canines from birth. We did manage to rear some of these, and a proud effort was made by a collie bitch I had at the time. She was rearing three puppies and was decent enough to allow two of the orphan pigs to share the puppies' upbringing. I took a picture of this situation at the time, and always meant to send it up to the Daily Mirror, Candid Camera or Private Eye.

Poultry-keeping is now so specialised that it is rarely one sees any hens on general farms; the fact is that hardly any eggs are produced free range. It amuses me to see the public stopping at farms to buy eggs under the illusion, presumably, that there are millions of unseen hens dashing around the fields, when, in fact, the eggs are likely to have come out of the battery.

I started farming at the beginning of the deep litter era. Hens were put into large houses or sheds, and provided with a litter of wood shavings and chaff, helped out, incidentally by me with a fork on occasions. This was supposed to keep the litter in good condition so the hens could live their ten months of active productive life in a happy egg-laying environment. Looking back, I think the system worked reasonably well.

A new theory at the time was the provision of electric light to fool the hens that winter was really summer to extend their laying period. This has become more sophisticated now, and consideration is given to light patterns in the actual rearing of the birds, to determine their egg-laying potential.

Two of the main hazards were cannibalism and egg-eating, both carried out by anti-social hens. Therefore, one kept a constant watch on the flock. When a culprit was identified, it was openly assasinated as a warning to the others.

During this period, I had one lad of eighteen to help, and to compensate for the low acreage of the farm, I needed to develop a few high-output enterprises.

Apart from the livestock, I had about four acres of potatoes and a few acres of cereals. I grew as much grass as possible, conserving the surplus into silage, because all my life I have hated haymaking, and good silage is the stuff from which milk is made. Strange, you may think, a farmer who hates haymaking, which is generally considered to be one of the 'fun' parts of farming life, but I was a member of the great army of hayfever sufferers. As a child I spent hours on hot June nights under a towel, inhaling the pungent fumes of Friars Balsam.

If there's anything worse than having an asthmatic or allergic condition from childhood, it is being unlucky enough to have an onset of this condition in middle age. In other words, if you get it early you can grow out of it, but if you get it later it seems to stay.

Anyway, I went in for silage, which is cut before the pollen stage, in preference to horrid dusty hay.

I had also dabbled in sugar beet, mangolds, and even parsnips, but had come round to the thought that my fifty acres was too small a unit for so many fiddling issues, and gradually concentrated on grass, even to the exclusion of cereals.

Poultry breeding had been reasonably successful, so shortly after my marriage, I decided to expand in this direction.

Chickens were bought in batches of one hundred, direct from the hatchery, and reared in a device known variously as a brooder, hoover or mother hen – all excellent devices for nurturing the young chick for the vital first weeks of its life.

When chickens are hatched they have enough food in them to last for about two days, although it is necessary for them to have water. This operation took very little space, but it was essential to find a location that was easily accessible and free from rats. An old loose box I had was useless because of this danger, so we decided that as there was only Marie and I in the house, a spare bedroom with plaster floor and jolly flowered wallpaper, would be an ideal site to rear an extra one hundred chickens.

It was too! The use of the bedroom was so effective that the fowls thrived with alacrity. There were slight disadvantages, of course. Marie had forbidden the carriage of buckets of chicken mash up and down the stair carpet, so there was the boring and strenuous job of climbing up a ladder, and entering the chicken room through the window. Then our affection for the poultry wore a little thin when they grew so fast, that we were awakened by the crowing of the precocious young cockerels each morning. What's more, they pecked off all the bloody wallpaper and reduced the floor level by 1 inch from the skirting board. The cost of a short-term cut was more than the job was worth.

Our main source of income was from the cows, which were housed and milked in the cowshed. When I first took over the farm, the existing cowsheds had wooden standings or partitions, and as the herd was attested for tuberculosis, the Milk and Dairies Order regulations for milking hygiene dictated that the old wooden partitions be replaced by concrete ones, and walls be rendered with concrete in one of the sheds. This seemed an unnecessary expense to me, and one of the sheds was left with the original wooden standings. It was quite in order to keep *dry* cows in that part, although I used to milk a few, pretending they were dry, before the lady from the Ministry of Agriculture came round to inspect them.

On reflection, it's amazing how much public money the authorities spent on this. The whole system probably changed a few years later, and the insides of such cowsheds ripped out to house the cows, not individually, but loose in the herd. In fact, the most recent innovation for dairy husbandry is to provide wooden cubicles in a big covered yard, just like the ones condemned many years ago as being unhygienic. In those days, however, the cows were milked where they lay, in contrast to today's practice of parlour extraction.

Loose housing of cattle could not be practiced if cows were allowed to grow their horns, and it is standard practice to disbud calves. This is a fine idea as I have strong recall of the day I decided to take the horns from my cows, with the vet's assistance, of course, by sawing them off.

There were two writers in the farming world whom I regarded as more or less idols as a boy. One was George Henderson who wrote that magnificent book, 'The Farming Ladder' about his achievements on a very small farm in Oxfordshire, and the other was the late lamented Arthur Street, well known as a broadcaster and writer of such great books as 'Strawberry Roan' and 'The Gentleman of the Party'. He also wrote a column in the Farmer's Weekly each week, managing to be controversial for all those years.

For many years, against prejudice, the latter had campaigned for at least two necessary improvements, the dehorning of all cattle, and the abolition of the four bushel sacks, in which farm labourers had to carry corn after harvest, and in particular, from the threshing machine. Well, both of these new ideas have been adopted; sacks have been abandoned almost totally as most grain is now handled in bulk, and few cows have horns today.

It is a comparatively easy job to remove the horns from a young calf, disbudding as we call it, but literally a terrible bloody operation to saw the horns of an adult cow. However, this was performed, with the result that the cows became more friendly, and laid together in the fields like sheep, with the fear of one another largely removed. Consequently, it was far easier to transport them in a lorry without them arriving at their destination frightened to death, or with their sides lacerated – a very human improvement.

I have always given my cows names, as it seems a bit impersonal to say, "Go and fetch number ninety-three up from the field, George." Each year as the heifers start to calve, I select a generic group from which to name them, and I do it in alphabetical order. For instance, last year I used countries, starting with Argentina, Belgium, China, Denmark, and so on. I have composers, Beethoven, Delius, etc. One year I selected football teams as the generic group, so started off with Arsenal, Chelsea, and so on. By this method it is easy to keep a record of the year they came into the herd, and to know immediately their age.

One of the great innovations of the Milk Marketing Board is the Artificial Insemination Service, which has raised the standard of cattle far above most expectations of years ago.

It was most amusing when recently I rang up for one of my 'Football Team' cows to be inseminated. The procedure is to give the cow's name and the breed of bull required. When asked "what's the name?", I replied, "Derby County." The man on the phone retorted "I don't suppose you've got a cow called Brian Clough, have you, as some of our lads would certainly love to see to that one!'

I thought this very unfair to Mr. Clough, who is, in my opinion, the best thing that ever happened to Derby County. He has done a great deal for the town, even more for the club, and is good value in his own right.

But back to my own tale. In 1958 I left Sinfin and moved onto the larger acreage of Scaddows Farm. What a palaver that was too!

Moving house is a big job, but moving farms is a gargantuan task. One has to milk the cows at the old farm, transport them in cattle lorries, with all the attendant chaos, get them used to pastures new, and milk them in an unfamiliar parlour. Imagine the even greater problems of a farmer's wife with three young babies at her heels. She bears the brunt of the domestic move. I was pre-occupied with the animals but Marie had the two-legged sort to deal with, plus chattels.

On the final load, one of the farm cats leapt out of the van and disappeared up a tree, never to be seen again. He must have liked the old place better than we thought. When the lorries arrived in the new paddock, the animals sniffed the air of Scaddows Farm with suspicion. They were released. A gate was open, and before we could stop a mass exodus, forty cows had disappeared into the next parish. Unlike the cat, the escapers were fortunately retrieved.

It wasn't the end of our trials and tribulations, either. On the Sunday morning, the milking machine broke down, and that was the day I was to make my debut on *What's My Line* on the telly. I very nearly didn't get to the studios, and if I hadn't, few people would have heard of Ted Moult, as reliability and punctuality seem more important than mediocrity.

I wouldn't dispute that in those early days, my world, and vision of it was very limited. I didn't want to see, and consequently didn't see, a deal beyond my own boundaries. This is a dangerous state of mind to get into, and I'm eternally grateful to the entertainment industry for coming along and booting me into a sense of perspective. Farming was – and still is – a form of loving slavery. You've got to love it, and you've got to be keen to stay the course. I'm hooked but my vision no longer ends at the fifty acre line.

Looking back through the record books, I note that in one year I made the magnificent sum of four hundred pounds profit. The hourly rate must have been a pittance. Notwithstanding, I regarded each hour away from the land as an hour lost. It's not the sort of job that one can cut short and go away at will for a long period. Continual work was becoming habitual until we went on a week's holiday to Wales. I had arranged for a friend to come in and milk the cows, and look after farm affairs in general. Workers in the towns and cities can down tools when the factory buzzer goes. Most of them never realise what a twenty-four hour job demands of a man.

As I've admitted, I became fanatical about my fifty acres. The only break I allowed myself in the working week was the Saturday afternoon game of rugger. In this state it's hard to disassociate the need to work. For my part I felt guilty when I wasn't hard at it. I poured mental scorn on those with soft jobs, the white-collar boys. I couldn't think that what they did was real work. I now realise that this was a puritanical attitude. By my high and mighty standards, people who did 'soft' jobs earned easy money. This perhaps was the primary reason that I invested the first 'easy money' I made from TV into buying more cows, new tractors and other *real* farm items. It salved my conscience.

This outlook was a wrong one, too narrow, too prejudiced. Whilst I do believe that a man is fortunate if he's doing a job of work that he likes, work is still work!

I suffer from not being able to enjoy idleness. If I'm taking it easy I keep asking myself, "why am I not working?", and can only feel at ease when I have actually done something. What an attitude for the era when we are told that leisure will be THE thing! I think that there is truth in the Victorian idea that work brings its own reward. They were forced to believe it as they needed to work so hard, but perhaps we will see the day when people with jobs (not measured work-day slaves on production lines, of course) will be the privileged ones and the last to go bonkers!

On the subject of prejudiced attitudes, women's liberation is well to the fore these modern days. I've got my views. I don't know if they're right. I don't know if they're wrong. One thing can be said, namely that a man in his endeavour to tame the forces of nature, and carve a career for himself, often tramples over the little woman. She can easily be excluded from social life, friendship, and be left in a lonely isolated position. The last person to realise it is her man. In a selfishness of which he is not aware, he sometimes imposes difficult conditions on the female of the species.

I wonder how many cases exist where a woman has helped a chap build up some kind of empire, working like hell and wearing herself out, only to be followed, after conking out or being discarded, by Mrs. No. 2, who has it easy ever after. The replacement usually has all the material benefits denied the first wife, if she doesn't mind being eventual nursemaid to an old bloke!

To quote my friend Graham Ross's favourite joke.

Wife on sickbed : "I don't mind you marrying again, John, but don't let her wear my clothes."

John : "Don't worry love, they'll not fit her!"

Yes – the Women's Liberation Movement has a case, but it has to be careful that its viewpoint doesn't become encased within narrow boundaries as mine nearly did.

3

Casual Labour

FROM the very beginning, I've regarded the work I do on television and radio as a form of casual labour, being grateful for the casual returns. It has meant a lot to me and still does, but farming has always been my first love. There are those people who think that I put on some form of act on television; they believe that the same is true of my natural voice on radio, but some folks swallow anything. I'm not an actor, and what they see and hear is the real Ted Moult. Such as it is!

How did it all start, then? Well, my wife Marie has great faith in her bloke's knowledge, and persuaded me to reply to a Radio Times announcement that said the B.B.C. were looking for people to take part in the Midlands and East Anglia section of the quiz competition, *What Do You Know*, the eventual winner claiming the title of 'Brain of Britain'.

After some hesitation I wrote, asking casually if they had any vacancies, and was amazed when some weeks later the B.B.C. replied, saying that if I presented myself at their offices in Carpenter Road, Birmingham, I would be given an audition.

Perhaps if the Birmingham Fat Stock Show at Bingley Hall, had have been another date that year, I would never have bothered, because I had a certain amount of trepidation about this audition. However, the dates of the two events coincided, and I decided to go along. After all, if the audition fizzled out, at least I could spend the day looking at the fat pigs and bullocks.

It was rather like being interviewed for a job. I was shown into an ante room, where one or two people were standing around taking a desultory look at the Radio Times, as we do in doctors' surgeries, when suddenly I was called into the inquisition chamber. A charming young lady, who I later knew as Anne Howells, immediately put me at ease with her very kindly manner, asked me a few questions, and at the end of the interview said "Your voice is a bit odd, but we've had worse".

Presumably they were scraping the barrel at that time, because I was selected for the first round, representing Midlands and East Anglia. The recording was held at the Samson and Hercules Hall, Norwich (that should give me strength, I thought), and I earned a £5 fee for my trouble. Four people took part. Two were to go on to the next round and two to be returned home without honour. I found myself in the latter category and came back home with a deflated ego, wondering what the hell had ever prompted me to apply in the first place. My radio career had indeed been short and bittersweet, it seemed.

Of course, it was a nerve racking experience, and the thought of exposing my total ignorance on the air made me feel like last week's lettuce – Wilted!

Some years ago I was invited to present the prizes at a school at Dalston, Cumberland, where the headmaster was one of my old teachers in Derbyshire. "What a grand thing," said Bill Green, "is television and radio for diffusing knowledge. Twenty years ago only I knew the extent of Moult's ignorance – now the whole world knows it."

There are performers who never appear nervous. I suppose some compensatory paralysis gives this illusion.

One of the people I met at Norwich was Franklin Engelmann, who was Chairman for the contest. His death in 1972 shocked us all. To describe a person as a "real gentleman" may sound like a cliché, but it certainly applied in Jingles case. I have never known a man more able to put a newcomer at his ease. He treated people on merit, whether they be Duke, dustman, camera assistant or higher rank, and above all he had earned everyone's respect through his own intellect and achievements.

It is perhaps an indication of my mentality, some would say a dustbin mentality, the essential quality for would be quiz performers, that I still remember some of those original questions.

A wide interest in diverse subjects, even at a superficial level, is more important for success in quiz programmes, whatever merit that may be, than a profound knowledge in a narrower field. I have always been interested in people, history, music and geography, and so on, and I've got the odd facility of remembering names when on occasions I don't really understand much more than the bare surface of what I am talking about. I would know for instance, that the pseudonym George Elliot really belonged to Mary Ann Evans, but perhaps not know that George Elliot had written Adam Bede – I never read Adam Bede.

I see a quiz person's brain rather like a library – lots of shelves, a few books in each and a very quick reference service without a section of depth in any particular subject. Being a little unkind to specialists, they seem to know more and more about less and less, until they know nothing about anything, or as an academic friend of mine once said "as the inner light brightens, the outside darkness becomes more intense".

Both Marie and I thoroughly enjoyed the trip to Norwich and we travelled back to the farm in our Ford Van, plus Terence Bond, one of the contestants, who had cadged a lift.

We had more or less forgotten about it until a few months later, a letter arrived from Joan Clarke, the producer, who had been asked to put on a television version of the show, called *Ask me Another*.

The central theme of the new programme was that the winners of *What Do You Know*, namely the 'BRAINS of Britain' would take part against scratch contestants made up of a team such as journalists one week, businessmen another, doctors, and – believe it or not – farmers.

The prospect of actually having my mug on television was even more terrifying than the recording in Norwich, so when I arrived at the Riverside Studios, Hammersmith, I was relieved to find my team mates, Ralph Whitlock a writer and farmer from Wiltshire, and a lady from Cornwall were in the same state of mind.

The gods certainly favoured the farmers that night. We had the run of the ball, which often determines victory or defeat in many contests, and I returned to Ticknall, experiencing in some minor way how the men of the winning team of the Cup Final must feel in their day of glory.

After this little episode, the producer Kenneth Milne Buckley, who, incidentally was married to Sylvia Peters, and became a good friend of ours, asked me if I fancied to be guest brain on a series of *Ask Me Another*. This led to the general illusion that I was a 'Brain of Britain'. It has always been an amusement to me that people assumed I won the Brain of Britain competition. I didn't. There were times when I felt like making this clear, either on the air, or on the telly, but soon came to the conclusion that when people had heard me for a few minutes it would be obvious that I'd been eliminated in the first round.

Of all the programmes in which I have taken part, *Ask me Another* probably gave me most pleasure. It was easy to do, the people were pleasant to work with, and there was none of the great pressures present in some highly financial shows which were popular at the time. No big prizes, but big fun, I thought.

Meeting people is fun too. Some of them stand out more clearly in memory than others, personalities like Anthony Carr who came from North Wales and had a computer-like brain. I wonder what's happened to him since those days? Another face that flashes before my mind is Arthur Maddocks. He was an intellectual bald-headed schoolteacher from Birkenhead, and would have surely won 'Pipe-Man of the Year' competition, had that contest been going at the time. He sported a special bent pipe with automatic sump or something, puthering out clouds of sulphuric fumes on almost every occasion.

There was Doctor Mostyn Lewis, Olive Stephens, Irene Thomas, Doctor Reginald Webster from Accrington.

There would be three people in the residents' team, versus various teams of visitors. On one occasion we even played a team of Quiz Masters including Gilbert Harding, Robert MacDermott and Jack Longland. Franklin Engelmann was the Chairman.

Quiz shows aren't as bad as the Spanish Inquisition. It was rare for anyone to score zero, nor can I remember anyone being humiliated. There was, however, one occasion in Manchester, concerning a young lady who had won a fantastic amount of money on an Independent Television programme, before the practice of giving ridiculously large amounts was stopped .

In the first half of the programme she answered everything, but in the second half not one answer did she get right. The Chairman confided to me afterwards that he had switched the questions round from one team to the other. I since wondered whether someone associated with her could have made a good career out of espionage. Who knows?

Ask Me Another was repeated the following year with a

different producer, a young man called Ned Sherrin, who became famous for *That Was The Week That Was* and is now established in the production of films and musicals.

The show travelled round the Country, into different regions. When we were at Bristol I met Ned's parents and his brother, and detected a certain affinity. His father and mother, Tom and Polly Sherrin were Somerset farmers, with cows like me.

People ask what effect my sudden projection on television had. One immediate effect was the receipt of many invitations to open garden fetes, fairs, and to attend innumerable dinners. What a prospect for a weight watcher!

Apart from a brief interlude on *What's My Line* I also received invitations for the odd T.V. show. There was one comical programme called *Laugh Line*, which was imported from America I think, and again produced by Ned Sherrin. This consisted of a representation of actors and actresses on a set, the contest being to re-arrange the actors and add a funnier tag line than the original. This is when I first met Pete Murray; I have almost forgotten the show and I am sure Pete would prefer to forget it also.

You know, the best way to demand attention, if not notoriety, is to be different. I never set out to achieve any kind of fame whatever, but I suppose that being a farmer that could answer questions, yet at the same time looking as though I knew 'nowt', was something out of the ordinary. One has only to look at contemporary figures to prove that whilst there are many people with ability, those who have some outstanding feature have a better chance of recognition than the run of the mill fellow. Consider Jimmy Hill's beard, Cyril Smith's volume, Brian Clough's outspokenness, Ken Dodd's hair and teeth, and Gerald Nabarro's moustache and ostentatious motor cars. These things have the facility to make their owner noticed, and then it's entirely up to the public to assess their true value.

Apparently I had that something different, because I was invited on various shows, including farming, where I met the producer of B.B.C. farming in Birmingham, the late Hilary Phillips.

In fact, I met many of the well-known entertainment characters of the time and still do today. Lots of exciting and interesting things happened, but sitting down to write this book, it's the unexpected that pushes to the forefront of my mind; things that might have been forgotten, but somehow are retained. Obviously they made an impact greater than I realised at the time.

I recall a producer saying to me in the early days, "Ted, are you religious?" I answered him by saying that I'd had a formal Anglican upbringing. Apart from that I shared, with everyone else, swarms of doubts. But having worked all my life close to nature, I thought that there must be something ordering the pattern of things.

"No, no," he interrupted. "What I mean is, would you agree to go on *What's my Line* on Sunday nights?" It struck me there and then, that that's one way of testing a man's religious beliefs. Will I work on Sunday as a casual labourer, or did I prefer to spend my Sunday nights in the pub? In those days, it was normal practice to send out T.V. shows live, if not lively, and not, as now, on tape.

What's my Line was a top-rated piece of entertainment, dating back to the days when horizontal hold first needed adjusting.

It will be interesting to see if the exhumation of the programme will be a success or not. Tastes and fashions change so, first dêjà vu, then hideous, then grotty, then quaint, and finally antique and admired again.

Today, television is criticised by increasing numbers for its content and quality. There are programmes galore and countless styles and tastes (or lack of). It's a sober thought that when the box was new, *What's my Line* was the highlight of the week for millions of viewers throughout Britain. It was probably responsible for creating the architrace television personality or pundit, and other similar species. I'm thinking now of the classic examples. Take for instance Barbara Kelly, David Nixon, Gilbert Harding, Lady Barnett, Jerry Desmond, Cyril Fletcher and numerous other top liners.

Like many programmes, *What's my Line* was an importation from America. I believe that in the eyes of the United States, but certainly not all the viewing public, the radio and television industry is one of their prestige-winning exports. I have a healthy admiration for that great country, but it is young and can hardly boast any culture of its own, excluding a few exceptions that prove the rule. Along came the entertainment industry: films, radio, television, and America grasped the opportunity to set a youthful pace. They are the pacemakers in programmes good and bad. Cast your mind over what's on offer. The amount of imported new culture from Uncle Sam occupies a huge chunk of our listening and viewing time.

I suppose that the Americans are compensating in this fashion for their obvious lack of self-generated cultural background. Sure, there are the Daniel Boones, Davy Crocketts, not to mention the Seventh Cavalry, all of which illustrates the terrific energy and growth that was precipitated at the birth and infant life of the World's currently most powerful nation.

Nearly all American Art is adapted European, probably created by first generation Americans, and showing the vigour of a first cross hybrid, to use an agricultural metaphor.

You may ask what of the men in the literary field, of the stature of Mark Twain? Who could fail to thrill at the lyrical "Hiawatha" by Longfellow? In the musical sphere there are men like Aaron Copeland and Ira Gershwin. Agreed, but the list is so short.

On my visits to the U.S.A., I've been struck by the pop culture over there. It's a two-way deal, I know. There's the unique example of the four boys from Liverpool who conquered the fifty states along with the rest of the world, but what impressed me was the similarity of the scene. It's largely an American export, especially when related to the commercial aspect. An advertisement on the box over there for spearmint, detergents, cornflakes, soft drinks, hard drinks and many, many other products runs along the same lines as the commercials we see in Britain, and it's us who are doing the 'imitating'! It proves how much we are influenced by our transatlantic neighbours.

I was telling you about *What's my Line* and how it was an adaptation of a similar programme imported from America. The right to perform it in this country was owned by Maurice Winnick, the ex-dance band leader. At the time that my phiz first graced (or disgraced) the telly screen, he had established himself as a theatrical agent, and at his request I went to see him at his office in London. He suggested that I appointed him to act for me. Theatrical agents were a new breed to me. I was used to cattle dealers, but not those who dealt in entertainment matters, so I said no. I've often wondered if my blunt refusal had anything to do with my short tenancy of a chair around the *What's my Line* table.

The first time I appeared on the programme, it was decided that lounge suits could be worn. Previously, it was a strict rule that the rig-out must be dinner jackets. Whether the powers that be thought that this Derbyshire farmer wouldn't own a dinner jacket, and made a slight concession to rustic inelegance, I'll never know, because the experiment lasted only the one programme, and dinner jackets again became obligatory.

I was therefore compelled to purchase a rather dashing new creation, not daring to reveal my own bashed up, beer stained garment, that had endured many hours of brutal treatment at low functions in my locality. Being parsimonious by nature, and also practical, I retained my old trousers, disreputable as they were, on the assumption that no one would notice their presence or their condition as I was sitting down. Anyway, if you keep an item of clothing long enough, it is sure to return to fashion!

Today's free and easy acceptance of any kind of apparel is a good thing, as far as I'm concerned. Farmers are not renowned for sartorial elegance, and the rigid code of dress demanded in former days must have been hard on the young people. I understand, for instance, that before the War, staff announcers for the B.B.C. always wore evening dress or lounge suits, depending on the time of day, excepting on Saturday mornings, when sports jackets were permitted. All this on sound radio, too! Maybe Lord Reith had a point, but in general I like the informal touch.

It's clearly imprinted on my mind that my debut on *What's my Line* was the day after I'd moved from Sinfin to my present address.

If many more incidents had have occurred during the removal, I certainly wouldn't have made the studios. As it was, when I got there, a suave young TV executive took one look at me and suggested that I clean the muck from my fingernails. He didn't like 'first-hand' evidence of cow dung being thrust under the noses of his celebrities. Quite right too. None of the celebrities would have appreciated it either, unless he happened to be Percy Thrower, I thought.

When the programme was underway, farming was never far from my mind. I made one big boob that first day. For those who've forgotten, or never knew the principle of *What's my Line*, the basic object was to guess the profession of the person who appeared before the panel as guest of the week. I asked this particular guy how I'd look wearing one of his products. As it turned out, it would probably have suited me, for he made bull-rings. I've still got the one he presented me with afterwards.

Another guest on a later programme was Gene Tunney, the boxer. As soon as I saw him I thought what a marvellous hand-milker he would have made. He'd got the biggest hands I'd ever seen and a grip like a steel vice – just the job for hard milkers!

I thought I was going to have some fun when I was invited to be the mystery celebrity on 17th January 1960. This was the spot on *What's my Line* where the panel blindfolded themselves and the celebrity disguised his voice when answering their questions. How wrong I was in thinking I had them fooled. Astute Lady Barnett had no difficulty in identifying Farmer Moult.

Once one becomes well known on television, invitations to appear on other programmes follow, almost as a matter of course, and some of these guest appearances of mine have approached the ludicrous.

Take for instance the time, about ten years ago, when I was asked to be the 'object' in a panel game. The idea was that a well known personality should do something entirely out of character, and the panel had to identify him or her. I was invited along to sing a song – behind a curtain. I took this quite seriously and learnt 'The Sentries' song from Act Two of Iolanthe, which isn't a bad song when sung properly. I even went to the extent of taking three lessons from a retired school-master at Ashby-de-la-Zouch, and reached a barely passable standard. Came the programme – live of course – and someone suddenly shouted, "You can't sing that song, Ted. The fifty year copyright taken out by Gilbert and Sullivan hasn't run out yet!" The best I could do was to quickly grab an old school song book, and struggle through 'The Ash Grove'. "Sung like a true Welshman," said Taff Jones, the A.I. man the next morning.

In fact my musical efforts on television were not appreciated at all, yet I've had no complaints from my cows at Ticknall. I took my mouth organ for a 'recital' on *Braden's Week*, and even my favourite T.V. critic, Nancy Banks-Smith of the Guardian, later wrote "He handled it with the dexterity of a man eating a ham sandwich."

There were saner moments, like *First Impressions*, with Drucilla Beyfus, Malcolm Muggeridge, Douglas Fairbanks Jnr., and Fife Robertson, or "Robbie" as he is generally known. Let me remind you of the details. A mystery guest would try to respond to various suitable remarks from a three person panel. For instance, I might remark "If I was to take a holiday, I'd take it at ....", and the Character would say Cyprus, or Skegness etc., and after a series of such remarks we had to guess the person to be identified. This programme enabled me to make a friend of Kenneth Horne and also Terence Brady, who alternated as Chairmen.

I enjoyed standing in for Jack Hargeaves in the Southern Television production *How*, but the distances involved and necessary rehearsals made it practically impossible for me to accept their invitation to do more of these.

There were the music programmes, which were my favourites – those where I was required to introduce the records.

The very first of these was *Desert Island Discs*, with Roy Plumley. I had only recently started doing odd programmes on television, when on my return home one evening, I was greeted with the message that a Mr. Plumley had rung. I immediately thought of Herbert Plumley, who supplied me with lime for the farm. Still, I didn't need to whitewash myself on the real Mr. Plumley's programme, and spent a happy introduction to the gramophone department of the B.B.C. selecting eight records for my imaginary abandonment on a desert island.

The morning I spent in the record library opened my eyes to the colossal range of recorded speech and music, carefully guarded by the Corporation. Such a comprehensive collection must become of inestimable value to successive generations of historians, when our efforts in the 20th Century are written up in retrospect.

I spent quite a bit of time later on, selecting discs for record programmes which, from time to time, I was asked to link.

Being an ordinary kind of bloke, there was one television appearance which didn't appeal to me much, though I suppose I could have been flattered. That was *This Is Your Life*, and Ted Moult was the subject.

I had travelled to London, ostensibly to do a pilot show for yet another quiz permutation, following which I was entertained at some considerable length in a restaurant, just opposite the television theatre where the final rehearsal was to take place.

Feeling well contented with life following my wining and dining, I just hadn't the guts, when Eamonn Andrews filled my sights, to suggest he took his book elsewhere. So I walked on the stage to be met by a Jersey cow, and also there were big bales of hay. "How much is all this worth per ton?" asked Eamonn. "Oh, I don't know," said I, "About £12 a ton", and lo and behold out came five kids, followed by my wife, who had been transported in strict secrecy all the way from home, for the occasion.

I had worked with Eamonn on a number of occasions, from the *What's My Line* days, and also on his own late night Sunday show, which was broadcast from Teddington.

I've always considered it easy to be interviewed, because the initiative rests on the man who is asking the questions, but I noticed some of the people I met on the *Eamon Andrews Show* were very shy and ill at ease, far more than I would expect, since they were from the world of entertainment. It occurred to me in some cases, but not all, that without a script, some of these people were absolutely tongue tied.

My children, of course, got quite used to seeing Dad on the telly, and I suppose they considered this quite a normal situation. On the rare occasions when they would perhaps relate some incident which had happened to them, concerning the broadcasting field, their enthusiasm sometimes was mistaken for showing off. When her class was asked in turn to state their hobbies, one of my daughters replied "Riding and going to television studios". The others called her a little prig! At least it taught her that the unusual or non-conformist in any society is always suspect, and that it pays to keep one's mouth shut sometimes.

They have been able to see how the TV industry operates, and Willie, Marie and I competed in the Southern Television programme *Ask The Family*, with John Ellison as questionmaster. We beat the Jack Watling family, but were knocked out by the Christine Trueman family.

Their main experiences lie in doing film inserts, and studio contributions for *Play School*. Now, there's a good programme for the kiddies, and thoroughly enjoyed by my own family.

Even Scaddows Farm has played a starring role. A complete colour studio turned up one day, on its way back from a by-election or the races and filmed us ploughing the fields and milking the cows, and also shot an interesting little cameo of the family strawberry picking, mending fences, and all the other jobs we accept as day to day routine.

Broadcasting generally has opened up a completely fresh field of vision for me, and provides a perfect foil for life on the farm, yet one factor that bothers me a little about television, is the ease with which the public at large accept the instant pundit. Someone like me, who was projected on the box, by luck, may be all too often blown up out of all proportion to real life.

What I have in mind, are those who assume an instant expertise and appear on telly giving opinions about everything from the world political situation to the nature of God. It's not their opinions that worry me; it's the way in which their views are accepted by a large number of people as gospel. Why should their views on the world carry more weight than *you* think about the subject? Most likely they don't. Listen to the other chap's angle, but stick to your own guns unless his argument wins you by sheer force of logic.

Accepting the other man's view is not easy. "A man convinced against his will, is of the same opinion still". Who said that? Off the cuff I'll try Goldsmith. If I'm wrong and you're right I'll give you a free punnet of strawberries, assuming I'm still growing them! In any case, most of us know the difference between right and wrong – the hard part is to tell between right and right!

4

The Gift of the Gab

I'VE NEVER thought of myself as having the gift of the gab. In fact, I wasn't divinely blessed with a ready tongue. Few are. Most, like me have to learn the hard way.

There's no denying that the ability to speak and present ideas in lucid form is invaluable, unless you happen to be a Trappist monk. It may only be a question of entertaining a few close friends in after-dinner conversation. Certain people do have a talent of making themselves understood without using words, but most of us need to speak a little, at least. On the other hand, some individuals, after uttering their conversation piece, still remain unintelligible. Still, it's an asset to speak well and hold the centre of interest. Certainly it is one of the ingredients of personality. With this in mind I want to tell you of some of my experiences, and pass on a few useful tips that may come in handy.

When I first appeared on TV I found it a nerve-racking experience. What's more, I still do! In my opinion, a person who can face the cameras without any reaction is someone to pity. He or she would have to be over-conceited, probably insensitive, and the result is hardly likely to make good television, nor gain high audience ratings if repeated. The only place for human Daleks on telly is in the Science Fiction series.

Experience has taught me that there is little difference as far as *nervousness* is concerned between being on TV or addressing a small or large gathering. However, one's reactions are not the only consideration. The thing that does change dramatically, according to each individual type of audience is the technique to be employed. There's no such thing as the successful one-off man who repeats a parrot-like script, no matter what the occasion, and irrespective of the size and nature of the audience. Anyone attempting a stereotyped performance is about as interesting as an unstrung puppet. I know. I made this foolish mistake once, but more about that later.

To make things plain, let us turn once more to television. When one first makes an appearance and is the focus of bright lights, there is a tendency to panic. Are you in for third degree, American style? Are there millions of hidden eyes looking at you nationwide? Few feel comfortable under these circumstances and until the 'victim' comes to terms with the experience, he feels like a specimen fly pinned on the proverbial wall.

The secret is to tell yourself that you're addressing a small family group. After all, this is true, for most of the viewers are made up of two or three people, plus maybe the kids and Fido the dog. You wouldn't mind chatting with them, would you? If you find yourself dragged screaming before the cameras, remember to regard it in the nature of a cosy chat with this one imaginary family.

There's another method that I've found even more relaxing, and that is to talk to the interviewer as though you and he were having a private discussion. Forget all about the viewing audience. Ignore the cameras. It's easier said than done, but in practice I've found that it works wonders (With apologies to a well-known ale). The best tip of all though, is to think about what you're trying to say, and not so much about the impression you're making.

Returning to the subject of varying techniques, it's obvious that to speak at the Albert Hall would require a completely different approach. The likelihood of a challenge of this size is remote as far as the average man is concerned, and that includes me. I don't think we need give much thought to it in these pages, except to emphasise the importance of acoustics. I make reference only because extremes clearly underline the major theme of this chapter.

After-dinner speeches fall half-way between the two extreme categories. You can see the audience, but they are small in number. What's more, by the time you rise to address them they are likely to be both receptive and sympathetic as a result of their wining and dining. They know they have to 'pay' for their 'supper' by listening to somebody spout, and it might as well be Joe Soap. Yet, even the most pleasant of audiences can get bored, and this is something that must be avoided at all costs by budding speakers. Here's a useful hint. I make a point of standing up and pausing for at least twenty seconds before uttering a word. This has two effects. It allows me to catch the eye of the audience and in some daft way it imparts a sense of authority. In other words, they are waiting for the speaker, not the other way round. It's a fatal flaw to leap up and blurt out. This betrays you as a rank amateur.

If experience counts for anything, I have been lucky enough to address audiences all over the British Isles, in many different settings.

All types of audiences? Well no, because to be an all-weather multi-purpose orator one must be able to control a crowd, listening to something that they do not want to hear, and no doubt disagree with anyway.

Not that I am comparing myself to a politician who may have some indigestible guff to pronounce, nor am I likely to invite the hostility of those present by trotting out a few innocent spontaneities on whatever subject seems relevant at the time.

They listen, presumably because they want to, or because something better is coming on later. If they don't or cannot listen, I'm sunk. Take one occasion at a local N.F.U. Dinner. There was a power cut, the soup was cold, the fish course disappeared, and the women thought they were sitting at the North Pole! Warmth is an essential ingredient of life, and just as you cannot teach kids to swim in cold water, you cannot expect a freezing bunch of captives to hang about while someone is droning on about pig prices. The lights went out, the mike went out, and so did some of the audience. I blew out the candle and followed suit!

There's nothing more disconcerting for a speaker to hear feet being shuffled and the low mumble of voices in the audience. Once that happens, the beginner is likely to lose control and end up like a spinning top, going off-balance. Another thing that's worth remembering is to address the audience in a loud, clear voice. Don't shout at them, but don't whisper either. Otherwise, Uncle Jack on the back row with his ear trumpet won't hear a word. Whispering or shouting certainly won't enhance your reputation, even if the subject matter is brilliant.

I've found that the most difficult engagements to fulfil with complete success is making a speech outdoors. There are so many hazards that I hesitate to list them. First and foremost, it may start to rain when you've just got under way, and inclement weather will deter the most enthusiastic of crowds. If it's a fete or some similar function, the audience is hardly likely to have wined and dined a matter of minutes before, so there is no softening-up process. Then there are the obvious distractions: things are happening around and about, in the sideshows, stalls, and various contests. Then little Billy will suddenly fall over and start bawling his head off! There are so many unexpected hazards that you can do little or nothing about.

Despite this, there is one important feature that should be watched from the very start: remember the acoustics. Usually they are terrible at an open-air event. Today there are firms specialising in sound equipment and they are well worthwhile employing, if the budget can stand the cost.

On one occasion I went to the Highlands of Scotland to open a Grand Garden Party. I have always been a bagpipe fan, and even tried my hand (and arms and lungs) on them once, at Gun Club Barracks in Hong Kong, of all places, when I was a guest of the Black Watch, but nowhere are they more effective than on the braes of their native land. Unfortunately, depending upon how you look at it, the bagpipe was all the assembled crowd *did* hear to any degree. Maybe they were lucky!

The worst experience I ever had at an outdoor function was when the amplification actually bounced back. This was a totally incapacitating experience! I heard my own inane remarks boomeranging towards me, seconds after I had uttered them! It was a bounce-back of the worst order. The pundits say that reputations are won and lost on the playing fields of Eton. I don't know about that, but I'm sure that a few reputations have been won and lost on many a field turned into the site of a church bazaar, political fund-raising function, or something similar.

I've headed this chapter the gift of the gab! Don't think that you have to be born with a special talent. I wasn't. I hope that I've made that plain. Develop your potential for speaking in public. Let me tell more about the first time I attended the Debating Society at School. It was an impromptu affair, and my name was drawn out of a hat. I was so ignorant of procedure at the time that I didn't even know how to correctly address the Chairman. There were about fifty people present, as I stumbled through to the best of my meagre ability. I was a washout! The lesson went home and I became determined that I would soon know how to address audiences large or small. At the time I had no idea that I would make professional use of the art, but even if my life had taken another direction there's no doubt that what I learnt that day would still have been of value, in one direction or another.

Anyway, I became very interested in debates, partly because of the challenge. I couldn't have done too badly, because after a short while I was made secretary of the School Debating Society. The Secretary is the one who does the work, whilst the chairman takes the glory. This resulted in my making sure of the Chairman's position when I joined the Young Farmers' Club years later. It worked too! As Chairman I did most of the spouting and got the kudos. The poor hardworking Secretary did the toil with little thanks. In a less light vein, I can say that my days of speechmaking as a member of the N.F.U. were of tremendous value. It soon became second nature to speak with reasonable effect, and if the audiences were not paying attention, at least they were not dropping off to sleep. Maybe they were thinking about frit fly or turnips or something.

If one wished to express an opinion, it was necessary to realise that as the hour of milking drew nigh, the committee members would start leaving the room, rather like the members of the orchestra in one of Haydn's symphonies. They would depart in ones and twos, not out of protest or boredom, but out of consideration for the swollen udders of their cows.

The year when I was Chairman of the National Farmers' Union, our local branch went to the Annual Conference at the Central Hall, Westminster. This, of course, was the big 'do' with a capital D. We were the chaps 'up from the country' visiting the metropolis. As I remember it, none of us had straw in our hair, but I'm sure that many Londoners still regarded visitors from the sticks in this way. Not now though, surely.

After each conference resolution there was a general debate, which was used to good effect by yours truly and I gained experience. This was important to me and stimulating. I'm a man who has always needed a little diversion from work, in fact we all need a hobby. If someone had told me at the time, that eventually I should be appearing on television and radio, I would have laughed at the thought but that's what having a wide-awake wife does for you. Watch it, chaps! You never know what you're going to be pushed into!

Taking a look around at contemporary speakers, those I admire most are Enoch Powell, Lord Soper, Baroness Stocks, Malcolm Muggeridge, and others of similar ilk. It's not because of any political persuasion, but due to the effectiveness of their oratory. They seem to make listening compulsory.

I'm not saying that I agree with all of their viewpoints, or certain philosophies. I'm weighing them now in the scales of ability as far as public speaking and presence goes.

My advice to budding speakers is to regard each occasion as worthy of an individual approach. Each to its own. In my own case, experience proved a good teacher. I'd already had an eye-opener, inasmuch as I'd no idea that so many organisations sought speakers of all shapes and sizes and with as many varying subjects, but the market being immense, is so demanding. In addition it has it's own style. It is fatal to rely on a standard lecture, or lectures. They just won't do. You may be master of your subject, be it oil painting or mountaineering, but the public is not as gullible as many like to think. They instinctively spot the speaker who is rolling out a stock after-dinner speech, particularly if they have heard it before.

No audience really wants to hear a string of risque, or even filthy jokes, without a clue as to what the invited lecturer thinks about the subject in hand. Not that I dislike jokes, whatever their colour, the only criterion is if they make me laugh. The most skilful, and consequently most demanded after dinner virtuosos rely on wit, not pickings from Playboy or Readers' Digest.

Sincerity is a *must* for successful public speakers. This is true beyond question, but I must emphasise that sincerity isn't the be all and end all. No virtue has an absolute quality. Hitler was sincere, whatever his faults, but it didn't make him a good man, or right in his convictions. In fact almost every tyrant since the world began has been sincere to his own cause. This 'virtue' magnifies the personality; magnifies and makes it magnetic. I remember reading the biography of Al Capone, the most infamous of Chicago gangsters. He was responsible for countless brutal street killings, both at first-hand and through mob rule, and his name was spoken with revulsion throughout the civilised world. Yet, when he was finally arrested, and sent to trial, what do you think his *sincere* reactions were? He was hurt! Just imagine that! His lawyers, and others who knew him during those days, swear that he went to the electric chair sincerely aggrieved. He considered himself a public benefactor, who had helped the poor, the repressed, and the old. If it had been necessary to eliminate certain people, it was for the good of the community, in Al Capone's eyes. If this gangster thought he was right, why try to tell any man he is wrong? However, it seems that Capone was sincere.

I'll end this piece on speechmaking by telling you of some of the *plus* benefits. After I became fairly well-known, Foyle's placed me on their lecture list. This I regarded as an honour, as well as being lucrative business, but I had to give too much time. I found lecture tours too demanding. Travel here, travel there. It just didn't fit in with my farm programme. Show-biz has always been casual business to me.

During my travels, however, I saw parts of the country that I would not normally have visited.

It is surprising to realise how often one puts oneself completely and trustingly in the hands of a stranger during these lecture and personal appearance tours. A company in the West of England had invited me to attend their firm's annual party, but I felt less than festive when my chaperon for the evening slid his car into the ditch on the way to yet another party. What really sticks in my mind, is that at the time he was discoursing on the possibility of spiritualism and contact with the dead – a bit previous I thought.

Also on these tours, one must face the unexpected, as I discovered when I visited a Yorkshire Literary Society. The audience had a great time (I think), except that all the speeches were so long that half the members and guests, who had imprudently supped too much ale, kept leaving the room. I wondered how many would be left to hear my ramblings, as I was number eleven on the batting list, so to speak. "How long would you like me to speak for?", I enquired of the President, as the clock approached eleven P.M. "Well, about an hour, lad", he replied, "We like our money's worth up 'ere, tha knows!'

This is about as off-putting to a speaker as when a kind host or hostess makes such a generous introduction as: "Well, we've heard some amusing stories from the last and previous speakers, but I want to introduce you now to a chap who will have you falling about in no time, and who . . ." This is no help if your effort goes down like a stone balloon.

My travels were not just confined to this country. Far from it. I did an American tour, amongst others, and found that they reacted to the chat in the same way as British audiences. Their problems are similar to ours. We're in life together, sharing hopes, joys, and disappointments. A problem is a problem in Boston, Massachusetts or Boston, Lincolnshire.

This makes me reflect on the Common Market. Although we are a member of the community, our ties with America are closer in the natural sense. We speak the same language and there remains to this day a special relationship, although dented as it is at times. This proves to me the power of a common tongue in the affairs of men, individual and as a community, and the importance of being able to clearly express oneself. John Bull and Uncle Sam already have the knack.

The European Parliament, despite its pretentions, can never be the same as the House of Commons, or any other talking-shop where people speak the same lingo. I noticed this when I attended a session at the United Nations in New York City. The delegates meant well, but were they understanding the word-for-word meaning of the other man? Only partly, I suspect.

I have always considered words to be basic aids to communication, perhaps not in themselves fully expressive without the nuances, the inflexions which should accompany them. In other words, it is the way that one says things that is paramount. If my wife asks me "Where did you get this meat?" for example, I would tell her the name of the butcher. If I ask her the same question she might well say "Why, what's wrong with it?"

Animals can make their thoughts clear with no words at all as any dog or cat lover will agree. This reminds me of an old gardener who was asked what date he thought was the best for sowing his peas? "Ah," said the old man, "Tis not the when that's important – 'tis the how."

On the question of English as an international language, there's no doubt that as the world shrinks due to jet aircraft and high-speed travel, the desire to communicate drives people to voice our mother tongue. It never fails to impress me that every other person in the civilised world seems to be able to speak some form of English. Poor General de Gaulle would probably turn in his grave at these words, but I think they are true.

Do you think that more people speak English than any other language? You do? Well, you are wrong. The answer is Mandarin Chinese. I know, since I made the same mistake in a quiz programme, so you see, some things stick after all!

English will become the leading international language, in my opinion, ousting French from the position. Why do I say this? Well, if you think of a Dutch aircraft landing in Japan, the ground-to-air talk will be in English. Do you get what I mean?

I reckon that's why we are so bad at foreign languages. We have an inbuilt arrogance, and in our hearts we know that time is on our side. When we do speak a foreign language we tend to do it with bad grace. We're the only nation who regards it as the duty of the other party to take the trouble to learn. On a recent overseas tour, a typical reactionary Britisher said to me "Why can't they speak English and drink gin and tonics like the rest of us?" That just about sums up the attitude.

One final point for embryo speakers. Never fear that you will run out of material. Life itself provides an inexhaustible supply of subject matter.

This brings me back home to the time I addressed a local Rotary Club at Swadlincote. They seemed to enjoy it and I was asked to speak a second time. When I replied that I'd got nothing left to say, they persisted in their invitation, and I found myself rising to the occasion on short notice. I gave a topical speech and one that went down very well. How did I do it? I picked my subjects out of the Sunday papers and then expressed my views on each of them. Life in action.

Knowing when to stop is more important than knowing where to start. Having suffered from bores at dinners, Speech Days, and even weddings, I try to leave the audience wanting a little bit more.

The time to worry is not when they are looking at their watches, but when they are shaking them to see if their chronometers have stopped.

It's the same with articles. Here's one I did on Dairy Queens:

"I cannot quite remember who said it, but it is worth repeating. "Half the money spent on advertising is wasted, but no-one knows which half."

How does this apply to the Dairy Queens? Are they advertising and if so, what? Is it milk, cheese, cows, the desirability of becoming a dairy farmer with the prospect of siring such a winsome maid, or what?

Since I live almost literally next door to the current Dairy Queen, I suppose the chairman in his ignorance assumed that I must know something of the actual competition and therefore am well-equipped to speak thereon.

Alas, although I have knocked about the milk publicity scene during the last ten years, my knowledge of this particular facet of it is rather limited. I could have written a piece entitled 'Dairy Maids I have known', but this might have been interpreted wrongly and would have been a mercifully brief article. And in any case, as I spend ten times as long in getting milk out of cows, as I do trying to persuade other members of society of the enormous, almost unrepeatable value of dairy products, I thought that the best way to approach the subject would be from the angle of an ordinary farmer.

Well, the first thing that strikes me about the contest is that the whole thing is genuine. All the girls have to have a connection with farming, if not actually resident on a farm, and must know something of the technique of cow management. I suppose that if a poor girl is perhaps needing a stimulant to pep up a dull conversation, a remark such as – "What do you consider is the optimum vacuum that one should have in one's pipeline?" would be of inestimable value to have up her sleeve! Unless of course, the interlocutor happened to be equally crackers on milking machines.

The girls who go in for the Dairy Queen contest are a different breed from the gorgeous glamour pusses who make up the circuits around our seaside resorts. I say circuits, because it appears that the aspirants for the other titles set about the job in a very business-like way, a bit like Show Jumping without the horses.

I once had to judge, along with two more persons, a Beauty Contest at a seaside swimming pool. The three of us resolved not to select anybody other than a genuine local lass, but to our surprise, all of the last eight were 'professionals'. We were fooled alright.

Well, the Dairy Queen competition is not like that at all. In fact, most of the eventual winners get there by accident, having been physically heaved on to the dance floor like reluctant Sabines by lusty young farmers at the preliminary knock-out heats.

Another unique feature about this event is that charm, personality, beauty and intellect are not subordinated to sex appeal. The girls are not required to parade in bikinis or to do anything which would offend the most bigoted puritan, that is if there are any these days! The idea is to select a nice wholesome lass with a tang of country air about her, a kind of girl-next-farm appeal.

At one time, the most profitable animal to have on the farm was the dairy cow. Probably this is still true, although the outlook for beef is much better. If you are a housewife and have to pay twice as much as you are used to paying, then perhaps you will not agree.

Beef costs a lot to produce as it takes seven or eight pounds of food to make one pound of meat. A pig only takes three pounds to produce one pound of pork, and chicken has a ratio of two to one. I think fish are even *more* efficient, so who knows, maybe salmon, trout and the like will change places with beef in the cost league.

However, I think that beef will always be the first choice for those who can afford it, and it is to be remembered that the majority of the calves will come from the dairy herd. Perhaps the calves will become more important than the milk, but milk is included in the daily diet of practically every man, woman and child in the United Kingdom.

It's a grand sight to see the Dairy Queen elegantly drinking the ever-present pinta, full cream, please note. She would not be seen dead with any of the powdered skimmed stuff, euphemistically called 'non-fat'. Perhaps it kids you that it is something to do with slimming!

So don't put your daughter on the stage Mrs. Worthington. But you may enter her for the Dairy Queen competition if she's good enough, and good luck to her!

5

Are Pets Pests?

FOREIGNERS say that the British are soft as far as pets are concerned, and frankly I agree. It's something else we share with the Americans. They, of course, take it a stage further than us, with mass commercialisation over a wide field. Everything from Beauty Parlours to Pets Cemeteries flourish all over the USA.

Pets have no place on a farm. Their habitat is in the homes of pet-lovers on the domestic hearth. To a farmer, and those who work close to the soil, an animal is an animal. It wasn't 'til I was asked to do a cat-food commercial that I realised how many pets there are in our mainly urban society. Nor had I the measure of what they meant to their owners.

Not only do they see Fido or Puss in an entirely different light to the professionals, but they are usually as blind to our attitude as I was to theirs. For example, I get a lot of letters asking for advice from parents whose young hopeful has an eye on a farming career. The initial correspondence nearly always contains the phrase 'he is very fond of animals'. If only the writers realised! This is the last qualification needed for a career on a farm.

Love of animals is a fine attitude, but it can be detrimental. It's far more necessary to realise the hard truth, a truth that quickly becomes apparent to anyone in farming or veterinary work, namely that animals have to be bought and sold like bales of cotton or copper wire. They are a commodity. The only difference is that they are one of nature's commodities and not a product of industry. People don't get soft about washing machines!

Even in this enlightened age, the average man or woman rarely faces up to the fate of animals when they stop giving milk, or when they stop working. The simple answer is that they go into a nitrogen/protein cycle and are eaten by other animals, including man. In nature, nothing is wasted. One of my daughter's ponies, with incurable foot fever, ended as edible material for someone else's pet. And why not? Taken from the smallest ant to the largest mammal, the cycle consists of living off some other form of life. Many people shun away from this undeniable fact. It's a question of drawing a fine distinction between sentimentality and hypocrisy.

In a queer way there's a parallel between this process and the case of humans who leave their organs for medical research. Transplants immediately spring into mind. This posthumous, and praiseworthy generosity has much to recommend it. After all, when a person dies, he is finished with his organs. What better use could be made of them than to pass them on to a fellow human being in dire need? I know that there are those with religious objections. I don't look at it that way myself. Nevertheless, it does put into question the literal interpretation of the resurrection theory, or bodily doctrine. Does this mean that on the 'Great Day' there will be thousands of people walking about with half their parts missing? Or am I being thick?

I was brought up in the days of shortages. Nothing had to be thrown away. What man couldn't eat, his dog could. In the war years and for a decade after, it was considered immoral to use humanly edible food to feed pets. Every source of edible food had to be churned through the natural pattern, and it was only when food was classified as unfit for human consumption that animals came into their own. Today, the picture has entirely changed in the western world. We're told that half the world is starving, but the developed countries seem to be living on the fat of the land, so much so that manufacturers of basic products are glad to find an ever-increasing market composed of the canine and feline elite.

Until my views underwent a certain change, I shared the arrogant opinion that a primary producer of food, or the primary producer of any other essential product, stood higher in society than the middleman, the chap who came later, and into this category would surely fall the pet-food manufacturers. This thinking was part of my 'fifty-acre mentality' that I described earlier. Today I regard it as a very crude assessment of people's worth. I used to think that miners, farmers, brewers, gas stove makers, and those who provided us with basic requirements were higher up the status ladder than people like actors, fruit machine operators, advertising agents, estate agents, publicans, white-collar workers and the rest. Now I realise that it's outdated and narrow to think in those terms. I might have got it the wrong way round completely. Everyone to their own forté.

Isn't it surprising how many complicated issues there are revolving around the subject of pets? Let's get directly back to them. People keep pets for their *own* pleasure, have no doubt about that. It's the two-legged animal that seeks the affection, and wants the pet for what the four-legged companion can give in this direction. Pet-owners believe themselves to be kind to animals, and I wouldn't be at all surprised if this chapter doesn't bring forth accusations of a brutal and cruel attitude. But before you get out your pens, let's examine the subject objectively.

When a kind cat-owner is feeding Felix a delicious meal from out of an attractively wrapped can, does he or she consider that they are unkind to whales? Most likely it never occurs to them, and if it did I doubt whether they'd shed tears for those large fishy beasts who are being swallowed in vast numbers annually, by hordes of hungry pets. What it really comes down to is that whales are relatively unattratcive, remote, and don't give affection. Pets do!

Yes – there's a lot of hypocrisy talked about animals. I've seen well-intentioned folk walking in the country, who stop and coo at Buttercup munching away in a field. They will lean over a gate and stroke the cow, then go home and eat its sister in the form of a juicy steak! And this is how it should be. Everything has its order in nature. When the wolf eats the lamb it's to the Glory of God. That's the wolf's job.

In an imperfect world it's unlikely that there will ever be an equal distribution of nature's bounty. I doubt if she intended it in the first place. In the light of this, perhaps it's not morally wrong to pamper our pets and make them into four-legged gourmets. It helps ease what otherwise might be a waste economy through overproduction; it provides new outlets to manufacturers, thus bolstering the sales of the primary producers of food stuffs. I've really got nothing against animals being kept in luxury, but they'll never be a substitute for human beings.

I am sure that it must be wrong to overfeed an animal while Old Aged Pensioners are perhaps suffering from a shortage of variety, if not a shortage of calories. The old lady or gent would certainly not fancy dog biscuits, but it's a bit rough on the pup to feed it into obesity, just because its owner hasn't a child, lover or companion on whom they can lavish affection.

Affection can go too far. I know a woman who has two Great Danes on her bed – double, of course. No doubt her husband is in the other bed with the cat!

Perish any thoughts that I am anti-dog. I'm not, especially on the farm. I have a smashing dog, Digger – that's her name. She is a Welsh sheepdog, by far the most intelligent breed in my experience. She doesn't require to be taken walks; needs no special food and eats anything; comes when I call; barks at strangers but doesn't bite them; doesn't kill cats, and best of all doesn't leap at people putting great muddy paws all over their clean clothes.

I bred her from another sheepdog called Sweep, which had failed its eleven plus in the sheepdog trials. Sweep was given to me by an old friend, Trefor Jones, who trained such animals in his spare time from his job as a District Officer in the Advisory Service.

If I had to choose, however, cats would get my vote before dogs. I regard the canine species in general, as syncophatic creatures that crave affection and attach themselves in a servile manner to an individual owner. Cats are more independent and live their own lives. If there's anything in the reincarnation business, I'd want to come back as a cat. Apart from the independent angle, cats are cleverer than dogs. Judging by these remarks I suppose I'm what could be called a cat person. The first time I ever heard this expression was when it was used by Katherine Whitehorn, in the Observer, a very bright lady who I met whilst on programmes for B.B.C. Manchester. In one of her articles she referred to 'cat people' and 'dog people'. *You* can judge which you are by whether you want to growl or spit at my words.

Whenever someone talks about animals, the subject of cruelty always arises. This is another question that can be countered with 'What is cruelty?' One interpretation of cruelty is portrayed in the anti-vivisection shop, near the Trafalgar Square end of Whitehall, where horrific pictures of animals are displayed for all to see. It reminded me of the first time I removed a dead mouse from a trap – the eyes sticking out like polished black ball bearings. Fear shows through the eyes of beasts as it does in humans. It's a difficult matter on which to sit in judgment, as is the colour problem in South Africa and other parts of the world. Most of us like the benefits that accrue from practices we abhor. It's alright if we keep our eyes shut tight.

Think of the lives that vaccination has saved. Does it really worry us that a few Rhesus monkeys from India have been sacrificed to make such progress possible? If it saves a few kids, who can reasonably object? Nature and evolution have provided man with the means to exploit the lower orders to his own advantage, but is it also his privilege? Every species is interested in its own survival, but in the process many lesser ones are bound to suffer. Man is hiding behind excuses if he doesn't admit that in order to reach the top of the pack, and stay there, he has brought certain species to the brink of extinction. In my opinion, however, man has a moral duty to see that this doesn't happen.

From what I've said you may think I'm a callous sort of chap. I'm not, but I am a realist. I hate cruelty in all its forms, but recognise the paradox. As a farmer, what do I do if a flock of destructive pigeons descends on my crops? Do I go out of business, or retaliate with a gun. I've shot pigeons was a vengeance. If those I hit dropped like a stone, I didn't mind. It was when a humble pigeon flapped away with a broken wing that my day was spoilt. The compromise solution is the one I have personally chosen – to get somebody else to do the dirty work. But does this ease my responsibility, or just my conscience? We're back to the 'save-the-animal' lovers who eat meat.

From my early days of farming, I remember a rustic savage who was typical of a certain class of countryman. At harvest-time such men would cut the corn and bale it into sheaves, which were eventually piled in the stack-yard. The following winter brought the big day that these men looked forward to with sporting relish.

What had happened in the intervening months was that the stacks had become the local Hilton Hotel for rats and mice around the farm. The owner was obliged by legal regulations to place wire netting at the bottom of the stack. When the time came to move the last few sheaves, the farmer would call upon his neighbours, as surrounding farmers lent each other men. Their number was swelled by casual labour of the type that eschew regular work and prefer a life of dust following the threshing machines.

There they would gather around the wire netting, a lethal circle armed with a motley collection of weapons. As the bottom sheaves were removed, out would come the rats. Hordes of them! And the slaughter that followed was spectacular. It was an event eagerly awaited after the months of hard work. These 'sportsmen' respected their victims enough to ensure that trouser legs were well tucked into Wellingtons. A rat up the trousers was something to be afeared by even the stoutest heart.

At one of these threshing sessions, I was amazed and appalled at the sadistic streak that rose to the surface in otherwise decent and respectable men. One fellow, who looked quite a gentle chap, harpooned a rat on the end of a pitch-fork, but not content with this roasted it over an engine boiler. It wouldn't surprise me if he returned home and made a fuss of the domestic dog. I'm sure he would have been astonished if anyone had accused him of being a cruel swine. Shakespeare in his wisdom said, "there's nothing right or wrong under the sun but thinking makes it so".

Returning to Ted Moult, how can I reconcile my unsentimental attitude to pets when I accept an invitation to do a commercial for a pet food? Well – in the first place, I don't mind cats, and have made this clear. In the second place, I've said there is a need for the widening of markets in Western society if we're all to have maximum prosperity. It follows from this that the sale from animal pet-foods does far more than feed them. It provides gainful employment for the primary supplier, right through to factory level, and encompasses the whole selling chain. In other words, whilst we are pampering our pets, we are also pampering people, too.

There's one kind of pet-keeping that I particularly dislike. I don't enjoy the sight of caged birds. A robin redbreast in a cage puts all Heaven in a rage. Caged birds are my own pet bias, but is it worse than caging guinea pigs, or even the domestic hen? Surely if it's right for one it's right for the other. Come to think of it, we humans live in cages of a sort, and that *does* worry me. It seems to be fast becoming the fate of many people to be pushed into a filing-cabinet style of living. The result – the escalating modern disease of neurosis. I'll expound on this subject in my chapter on factory farming, but it's worth reflecting that people over-react about battery hens, whilst ignoring the human counterpart in urban society.

Let's move to the ultimate. Is the killing of cattle a wicked act? If they are slaughtered according to the rules laid down, I think it's the least cruel way of ending their lives. It's no use blaming men when Nature itself made the laws, although as her highest species it becomes man's duty to rise above the basic instincts, as far as this is possible or desirable. To be civilised we have to subdue some of our natural inclinations. If we did everything on impulse in the name of freedom, or by desire, there would be no civilisation and no social order worthy of the name. Man is sensuous and greedy at rock bottom. He has to call on his higher instincts to rise above lower appetites.

I'll end this discourse on pets by going back to my early days. As a child I lived in a town and had a dog, which became a substitute pal. I had plenty of human ones, but the dog was my best friend. When it died I was distressed and discouraged. I longed for its replacement. Like any typical owner I wanted to renew the affection that had been snatched from me by death.

As I got older cats became my favourites. This directly opposed my mother's taste. She disliked the feline and wouldn't have a cat in the house. In the end we compromised for a miserable goldfish, but you can't get much affection from those clammy creatures.

I turned to breeding guinea pigs in a large Tate & Lyle sugar box and got so enthusiastic that I entered them in Fur and Feather Shows. I can recall going on a Trent bus with my pal to just such an event. Our excitement knew no bounds, and we reached a high pitch of excitement whilst awaiting the judges' verdict.

The most entertaining of these affairs were those organised by miners' clubs. These working men had a real zest for living, which found its competitive outlet in the care and attention they lavished on their hobby, the showing of livestock of all varieties, including Belgian hares, Dutch rabbits, and the like. We enjoyed the Shows and the food that followed. For two eager and hungry boys, the highlight of the day was the refreshment provided. Sixpence each and eat and drink your fill.

My enthusiasm for exhibiting guinea pigs grew and grew. When I couldn't attend a Show personally I sent my live entry in a box by train. Box and entrant were later returned. This went on for some time, culminating in the top event of the year – the Fur and Feather Show at Bradford. I railed off my guinea pig, used to train travel by this time. When the event was over the box came back, but to my utter dismay was empty. This made me appreciate that it was debatable to use animals as a form of self-indulgence in order to win awards at Shows, especially if you finish up empty handed.

Indeed, it taught me the futility of showing animals for fancy, from the tiniest creature up to horses and fat cattle. Monstrosities have resulted from man's vain striving. Mink of special colour, hideous dogs, artificial Dalek-type animals. The catalogue is endless and must be an offence to creation.

I hope my words won't upset well-intentioned people, many of whom don't see things in my light. I feel genuinely sorry for folk who need gratification at second-hand.

If everyone shared my views on exhibitions and agricultural shows, I suppose there would soon be no such events to visit. Who would miss them if they packed up? The ordinary chap in the country looked forward to the local show in the old days, but with so many counter attractions now, the main support is from the city dweller. He, no doubt fosters in his bosom that old illusion of happy peasant life, where decisions are made by such individuals as himself. This is also in contrast to the inflexible rigidity that is the consequence of the industrial revolution, and by which system the majority live and work.

These shows are usually held in beautiful surroundings, and with the additional spectacles in the Main Ring, plus the traditional gymkhana events, have developed into a countryside diversion for some and an outdoor party for others.

The Royal Show at Stoneleigh is probably the best of its type in the World and I believe farmers appreciate the field trials and practical demonstrations which have become an important feature.

In the main, showing of animals is carried out by a minority group of professional breeders, and has become rather specialised, rather like show jumping or professional tennis.

Having had my say, I freely admit that the older I get the more tolerant I become. It's sad that many people who, like all of us, need the love of another, should have to find this by the devotion or cupboard love of a cat or dog, which shows that substitutes are needed in the home as much as on the football field!

My views on pets and allied matters expressed in this chapter are not a creed. They are just my opinions. If people want to enjoy themselves by overfussing their pets, who am I to criticise? Nobody; but I still reckon that the word is often spelt the wrong way. P.E.T.S. should be P.E.S.T.S. !

6

What a Bloody Subject

AS A countryman cum broadcaster, I'm expected to have opinions on all controversial subjects connected with rural matters, and one that's always coming to the fore is the vexed question of blood sports. It provides a good starting point for sport as I see it, cruel or otherwise, and allied topics in general.

In November, as the mists swirl round the coverts, the fox must look to its safety as the considerable resources of man, horse, hound, Landrover and foot follower, combine to chase him to destruction – and the militant, poster carrying antis make their attempt to save him from execution.

I am amazed by the strength of opinion against fox hunting, held by so many people in possession of limited information, and yet those in favour of this pastime justify it on moral, economic or other irrelevant grounds.

To me, hunting has one thing in common with writing poetry – its adherents do it strictly for pleasure. They enjoy riding across the countryside on horseback, seeing a pack of hounds work against the skill of a fox in preserving himself.

Hunting people emphasise that the fox is a cruel animal. True, but then the whole of nature is cruel. The viciousness of our bushy tailed friend (or foe) simply underlines the statement. To say they are helping the farmer to rid himself of a pest is nonsense. I don't consider foxes as serious pests these days. Indeed, they may be an asset as they kill rats and rabbits, which do far more damage.

In the days when poultry was kept outside on a farm, the situation was different. Foxes knocked a few chickens off regularly, but they must miss their chicken supreme these days, as very few hens have to face the rigours of an outdoor existence. True, they may eat the odd weakly lamb, but I suspect the lambs taken are either dead or near dead, anyway.

Actually, at one time hunting enthusiasts took great pains themselves to *minimise* the vices of Reynard, in an attempt to discourage farmers from shooting foxes. They also deplore the idea of any gamekeeper or pheasant rearer reducing the fox population in protection of *their* sort of game.

If my opinions are wrong, and the fox *is* truly a pest, there may be some truth in the argument that hunting is better than trapping, shooting, or other forms of destruction, on humane grounds. Yet, would this argument apply to deer. These are a terrible nuisance, although delightful to see – on someone else's land, of course. In my district I would say that deer cause one hundred times the damage that foxes do. Surely, the same logic should be applied by the hunting fraternity, but I doubt it. You can't eat venison if it has been mauled by hounds!

Whatever merit there is in hunting as a means of limiting or culling any species of animal, it needs no more than one or two riders, as neither they nor the horses do the killing. What we farmers object to is the arrogance of those who assume that we should be delighted to see posses of riders charging all over our land, and scores of foot-followers standing on, and straining wire fences, as their necks are craned towards the action.

Now this poses another point in favour of the pastime. Hunting creates employment in the countryside – even if it is only mending fences.

Hunt members however, realise that they are on a knife edge of toleration, and to give them their due, they are careful to minimise annoyance to farmers, particularly in shutting gates, and educating followers on the difference between turf and young growing cereals. Motorways, disease control on farms and intensive cropping are also factors which make life for the huntsman more difficult. In the mid 18th century, which I would imagine to have been the golden era of fox hunting, there were great areas of unenclosed land which allowed unimpeded gallops for miles on end. Problems of trespass, shutting gates or hostilities from antis, as exist today, would have been unthinkable.

Still, if people have this desire for hunting, perhaps it is better to chase foxes than people with a pack of bloodhounds. Perhaps even this will become popular (killing optional), if the fox makes a general retreat amongst the suburban dustbins, which are fast becoming a natural habitat.

In my chapter on pets, I made the observation at length that people only see what they want to see. They brand vivisection as cruel, but sublimely overlook the killing that takes place to feed their four-legged favourite. A similar mental blindness applies to wild animals.

When I kept poultry, I housed them at night in a wooden hut. It had a slatted-floor of the type in common use for hygiene purposes. Brer Fox, however, failed to appreciate my reasons and managed to worm his way between the ground and the floor. Next morning I was faced with the sight of appalling slaughter. From a total of one hundred fowl about twenty had been smothered, fifty had had their legs dragged through the slatted railings and bitten off, and there remained a huddled bunch of terrified survivors.

It goes without saying that there was no organised crowd of do-gooders that weekend, waving placards, and holding protests in nearby country lanes on behalf of the fowl population, and against the murderous fox. Sometimes our double-standards approach the laughable.

If today's opponents of blood sports could be transported back in time about 200 years, their particular 'soapbox' would certainly become a full time occupation. In those days, many of the most popular activities like cock fighting, bear baiting, even prize fighting had more blood about them than sport, and I, for one, am glad that the blatantly murderous pastimes are not part of our own way of life. One such diversion was called goose-riding, in which a live bird was supended by its feet with its head and neck covered in grease. The 'fun' part consisted of the contestants' efforts to pull the creature's head off as they galloped by.

Many of these cruel diversions were probably perpetuated by the betting that went with them. Nowadays, the telly provides us with blood and gore in profusion (particularly the Newsreels), and there is every opportunity to gamble any spare cash on football matches we don't watch, or horse races in which the fancy is selected from a newspaper. We may feel revulsion at these old 'sports', but are we in any position to judge?

I've never been over keen on horse racing, whatever has been running, and I wonder why it is that horse racing would attract but few spectators if there was no Tote or bookmakers, yet motor racing exists without any organised betting at all.

Perhaps the reason lies in the fact that the car is a more familiar possession to the general public than a horse, and while there are certainly a number of horsy people who love racing for its own sake, the man next door can more easily identify himself with racing drivers. However, even these seem to lose their personalities as they disappear into the tailored cockpits of their ultra sophisticated machines. When we could see the old time drivers struggling with their picturesque Maseratis, with leather straps round the bonnets, they did appear as men, and were not nearly as remote as today's drivers.

It seems a pity all sport has seen such a technical improvement in the constant search for performance, that the spectator must be *au fait* himself, to get value from his ticket.

Personally, I like ball games, but have always been an undistinguished trier, without much success. At school I played soccer and cricket, and developed a great admiration for centre forwards. When I was put in that position the ball too often hit the back of my head as I was usually facing the wrong way. Turning at speed was not my best achievement, and my ambition to get into one of the school teams was unfulfilled. I nearly managed it though, during an epidemic of measles. "At last", I thought, "This time they can't overlook Moult's talents". Imagine my disappointment when I read on the Notice Board that the match was cancelled.

Football today is undoubtedly an extension of show business, and as such, must have the attendant personalities. In my district, the best things to happen for a long, long time are Brian Clough and Peter Taylor, who have put Derby County back on the map. I was a keen supporter in the days of Derby's former glory, in the late forties, when Raich Carter and Tommy Doherty were the big Saturday draw among other stalwarts, but until recent years, there was a long period of mediocrity.

I did actually play for Nottingham Forest, and quite recently too, but this was in a round of *Quiz Ball* on television. Derby County were still in Division 2, at the time, and as the contest was limited to First Division teams, Forest were the nearest alternative.

Do you know why Arsenal wear red shirts? Because many years ago Nottingham Forest sent a set of their shirts to a crowd of Nottingham men who were working at the old Woolwich Arsenal.

Just another example of my dustbin mentality!

It has been my pleasure to meet some very interesting and intelligent chaps in the football set up. Bertie Mee, for instance, who worked in a local hospital at Etwall, Derek Dougan, obviously destined for a brilliant post-playing career, and John Osborne from Chesterfield. Chesterfield seem to breed good goalkeepers – take Gordon Banks, for instance.

Rugby was more suitable to my physical capabilities, as I've mentioned before. It was possible to take part in this game at low level, without too much skill, and I thoroughly enjoyed the alternate weekly bus trip to away matches. To my mind this was an important part of the game, and later, when the team travelled in twos and threes in private cars, I felt that the essential team spirit had somewhat disappeared.

I played my last game against Keele University – well almost my last game, as in April 1973, I was fool enough to play in a charity match at my local club – 17 years after I had had hung up my boots. Our team consisted of 15 assorted players, including one or two has-beens like myself, and we faced twenty females of assorted shapes and sizes. All very exciting, especially during one scrum, when a bra came out instead of the ball.

Boxing is another sport I find fascinating, but only because it occasionally throws up engaging characters like Henry Cooper or Muhammed Ali. Unlike golf, the man who does the most hitting generally wins, but I still hesitate to call it a sport, as to me it seems to be run by big commercial interests, exploiting violence for their own ends.

Of course, boxing has had its troubles from anti-blood sport groups, but even then, not to the degree of antagonism that is shown to hunting. Many people who are anti-hunting have no qualms when it comes to catching the timid rabbit, the hare, or the fish in the stream. Foxhunting is different, they quickly point out. Well, if they mean that you can't eat a fox, I suppose it's as good an excuse as any. Oscar Wilde would certainly have agreed with them. On the subject of fox-hunting, he described the quarry as the 'uneatable pursued by the unspeakable'.

My attitude to shooting is as selfish and complex as the next man's. Maybe the difference is that I recognise the dilemma. If pigeons were eating my brussels, I would shoot them or have them shot all day long. Why not? That's much more defensible than the motives of people who rear pheasants for the purpose of gunning them down as game in the Autumn. In similar vein, shooting pigeons to protect crops is quite a different proposition to the type of so-called sportsman who takes potshots at tame pigeons released from a trap. This is not a popular activity today, and a good thing too! Clay pigeon shooting has taken its place – a civilised alternative. There's an uncomfortable parallel however, between the rearing of pheasants to hunt them down later, and housing pigeons to release and kill. One is only a step removed from the other in my eyes.

If I give the impression that I'm against the shooting man, let me correct it at once. I've used him as a sort of Aunt Sally, to let us see our own faults, warts and all. Frankly, I think that the hunting, shooting and fishing types generally appreciate the countryside more than most. There's no doubt that their opponents know much less about it when put to the test, and probably see little of nature in its own setting.

It's not death, but life that we've got to think about. The preservation of the species must be our aim. To put this plainly, where any activity of man reduces the numbers of a given animal or bird to danger level, action should be taken. Paradoxically game keepers don't help in this connection. Their job is not to preserve the species as a whole, but rather to ensure the propagation and protection of the particular prey that their employers like to hunt or shoot.

The old saying 'there's more than one way to skin a rabbit' can be applied to nature in general. This was brought home to me when I met a couple named Arthur and Margaret, who live in Leeds. They enjoy 'hunting', but with a difference. In common parlance they could be termed bird-watchers, or more correctly ornithologists. I wasn't long in their company before I came to appreciate that people who watch get far more pleasure than people who shoot the creatures. Arthur and Margaret identified birds to me that I didn't know existed.

To sum up, it pays to sit back and consider every aspect of a situation before arriving at a judgment. We can kid ourselves that we're dedicated and firm in our views on controversial issues like blood sport, yet be burdened with equal guilts that are apparent to others. There's none so blind as those who don't want to see.

7

Community Relations

MANY, many words have been spouted about the Common Market, some sensible and informative and some a load of rubbish and as a farmer I have been asked many questions on the subject, so here is the situation as I see it.

The thoughts of most people are concerned mainly with what they can themselves feel, like eating, sex, weather and paying taxes, so it is not surprising that the loftier aims and objects of the E.E.C. are disregarded. In fact the price of butter or even the selling of small mountains of stuff to the Russians are not the main fact in the creation of this mental block.

Sure, the new high food price system took the country by surprise, due to external factors. What should have been achieved over five years came almost overnight in a rush, and no one in their right mind can ignore these price rises. I got an abusive letter from some old dear in Shepherds Bush, objecting to what someone else expressed on *Any Questions* about O.A.Ps. having enough cash for their meat and two veg. It was someone else who sounded like me – poor chap! As I was abroad at the time of the broadcast I felt somewhat hurt, so I replied suggesting that the author of the letter should listen more carefully. I didn't bother to find out with whom the lady was confusing me, but it just shows how inaccurate reported speech is!

In case there is anyone around who does not know the set-up, this is how I see it. The Common Agricultural Policy (or CAP for short) is designed for keeping food prices high enough to provide a living for food producers of the main farm products in the member countries. Under CAP the farmer gets the whole of his return from the market price within the community, or from subsidies which will be paid to farmers who export goods outside the nine country community.

If the price falls below an agreed level, called the target price, the men in Brussels will intervene at the 'Intervention price' and take the produce via their agencies, for whatever use is available. Maybe for re-exporting or in some circumstances dumping.

To prevent any cheap food arriving at Common Market ports and undercutting home prices, the authorities fix a 'threshold price' in order to calculate a levy which the importer must pay into the Common Fund. This is important to Britain as we do import a large part of our food and consequently have these levies.

The Common Fund is then supposed to return the cash needed for regional improvement policies.

In other words, we in Britain have reverted to a system of protected agriculture, similar to that existing before the Repeal of the Corn Laws in 1846.

Free trade will be possible within the community and an interesting fact is that this is potentially a bigger market than the whole of the world in the middle of the last century.

The advantages of this mass market, we are told, will make it financially easier for the average wage earner to shell out for his grub, but as such developments take time, it was curious to see our chaps in Luxembourg pressing for a cheaper food policy as soon as they got past the door. It rather appeared as if the newest members of the club were trying to change the rules as soon as they got in. Let's face it, the cost of food will be high. Even if world prices become lower (and who knows?), the housewife will not benefit, because the British Levy Bill payable to Brussels will increase accordingly.

Possibly some subsidy will be thought desirable, as the days of buying up cheap parcels of food from all over the globe are surely over. The parsimonious policy of keeping British farming land run down and producing at half cock, has been proved unjustifiable.

We shall have to get used to paying more for our daily bread, but whether we pay out of pocket or taxes is still an open question.

A possible risk to the continuance of CAP are surpluses, which could occur through increased technology in food production.

Suppose for instance, through the encouragement of reasonably high prices, production of food within the Common Market increased, at the same time the U.S.A. brought into production the idle acres of the 'Land Bank', and for once things went right 'down on the farm' for Russia, the situation might be as follows.

With the Russian market temporarily closed, North America would be pressing to sell her cereals in the European market but would be unable to get over the Tariff barriers, as a result of the surplus of high cost home grown produce. Under these circumstances the CAP would 'bust'.

Over the last twenty years British farmers have bitterly complained about dumping, with real justification. Now that we are a member of the E.E.C. we must be careful not to participate in this terrible practice. It is only acceptable to dump food on the really desperate, to avoid starvation or similar emergencies.

The butter mountain is a good case in point. It seems to rise, become embarrassing, and then due to drought and a policy of Dr. Mansholt to shoot some continental cows, melt away again. Unfortunately, cows also produce beef, which also becomes in short supply. So up again comes the greasy Matterhorn until the officials are wondering what to do with it.

They can't dump without upsetting countries like New Zealand, so they export at a knock down price, thus enraging the old age pensioners.

What needs to be explained is that these seemingly crazy operations are unavoidable. To ensure a regularity of supplies, one must pay a premium. Wild fluctuation in food production is just not possible. Take eggs. If overproduced, they become so cheap that the producers lose money. Result – they go broke or out of business. Result – next year's eggs cost the earth and consumers go without their omelettes.

The Potato Marketing Board have developed a very successful scheme to combat the danger of surpluses. The acreage grown is controlled by a quota system, hoping that there will be a small surplus only, which may be fed to animals or even destroyed. In the main, this works well, providing a regular supply of spuds at reasonable cost.

The consumer public becomes irate at the thought of skim milk being poured down disused Welsh coal mines. However, this is a drop in the ocean compared with the total milk production. Nor do the public realise that in most cases, the cost of rendering fit for storage (drying etc.), plus transport and keeping, may make any other alternative unfeasible.

The fact is that the production of food is not an exact science. It takes a long time to plan, it is subject to the vagaries of the weather, and it is often perishable. To have cheap food on a free market there must be an excess occasionally. Farming is an industry like any other and one cannot imagine a manufacturer blindly producing more goods than he can sell. Farmers are the same, seeking the best return for outlay and effort. At least, farmers cannot be accused of laying off staff, due to overproduction. Restrictive practices are not unknown in industry. They have a place, but only after general rules of productivity have been applied, I would think.

Returning to the Common Market, I believe that in spite of the initial difficulties, the country has taken a step forward. Forget for a minute visions of bureaucratic manipulating of the regulations to suit individual advantages, and the representatives' pragmatic approach, and readiness to change the rules as soon as one member finds the situation too difficult to bear. Instead consider the idea that economic strength comes from larger units. I'm sure you will agree that this idea is valid.

Bismarck cemented Germany with a Zollverein or Customs Union as a first step towards political union. The States of America would never have achieved such progress unless they had united, although the country suffered great harship in the process. The great issue of the American Civil War was whether to split or not to split, and not slavery as popularly imagined.

These two examples did, at least, have the same culture and language, so it is a hope for the long term, that Europe will become not only a collection of members of an economic league, but an area of true authority. The price of margarine is therefore a side issue of small moment. The true aim is one of sovereignty.

It's a bit like marriage. No one knows the benefits, or the cost, without first giving up his freedom. I suppose its success depends on how the partner is functioning too.

Europe has much to offer the world and has indeed, a debt to much of it. As far as I can see it, if Europe is to take its place as one of the five centres of the First Division, i.e. U.S.A., Japan, Russia and China, it must plough resolutely forward, in the framework of the E.E.C.

I get the impression that Europe is irritating the U.S.A. by too slowly emerging as a solid political union. The Americans forget how long it took them to become united, and Uncle Sam will not remain the provider of credit and resources, for ever, whether for reasons of self-interest or altruism.

Indeed, that country, shaken by the realisation of its trading position, dependence on foreign energy resources, dollar troubles and bewilderment that its disastrous military adventure in S.E. Asia was so unappreciated, and also suffering from the trauma of washing so much dirty linen in the Watergate affair in public, might well take a new bearing on a more nationalistic course.

Isolation of our own Continent cannot be envisaged, as the world is so small that trouble or distress in any part can have effect on the rest of it. But Europe must definitely stand on its own two feet.

8

The Roast Beef of Old England

EVERYONE everywhere has an opinion. The lower the subject, the higher the number of self-styled experts ready to express their views. Whilst most people shut their eyes to problems of a psychological nature, nearly everybody is ready to expound on daily matters such as the weather. Most popular of all is the stomach. Go into any bar and start talking about religion or politics, and people don't want to know. Say what a lovely day it's been and what a lovely meal you had at such and such a hotel and you'll be the centre of attention. The secret of success is to stick to the basics. Be like the animals – think and talk about food, shelter and sex.

Anyway, I'm certainly talking about food in this chapter because it's such an issue of the day. So too, are mortgages, which cause quite a stir, but still take second place to the tummy. When something happens to food, or the price of it, there's a big uproar from everyone. It's not only the Common Market that causes prices to rise. Take beef. The roast beef of Old England is more than food. It's an institution that arouses in many breasts the same sort of emotion as Hearts of Oak or John Bull. In short, it's a tradition that has assumed a kind of reverence. In our insular way we consider ourselves superior to the French with their frogs' legs, snails and edible doormice. We reserve an equal air of superiority over the German with his taste for black bread, sauerkraut and Frankfurter sausage. In fact, the British are patronising to the whole of the human race as far as their weird choice of foods go. Natives of these islands cover up their culinary inadequacies by asserting that plain straight forward food doesn't have to be disguised with sauces. We tend to imply that the other man garnishes inferior raw material to make it acceptable. Such critics should remember that one has to be an expert to make poor stuff into good stuff. Experience teaches, and it's a clever cook that turns a low priced product into a delicious dish.

I was invited one night to dinner by two friends who had spent years in Brazil. At the table I found myself in the company of another guest whom I knew to be a local butcher. Our hostess served up a delicious dish called Feshuada, which was enjoyed by all so much, that the butcher asked for another helping. I've never forgotten his expression when he found a pig's ear in the stew pot and learned that this was the principal ingredient. There's nothing wrong in that, of course. It's just a matter of taste.

In this country we have always assumed that we eat the best and the finest meat at the lowest possible prices, a sort of divine right. It has therefore come as a shock to Mr. and Mrs. John Bull to find their staple diets under economic pressure, with meat in the forefront. The horror of it! Food isn't as cheap as it was!

Although our housewives howl like mad, the fact is that inflation of food prices over the past few years has been slower than wage rises. In 1962 it took an average worker twenty-eight and a half minutes to earn twelve eggs. Ten years later, he could earn twelve eggs in half the time. People easily forget what they want to forget. Along with more affluence, comes more eating out for pleasure, and in excess of nutritional requirements. This creates demand. People are simply living better than ever before, and are able to afford what would have seemed like unashamed luxury twenty or more years ago. When, as a result, beef prices go up it seems an affront to national pride, and housewives throughout the land can be heard loudly complaining that money doesn't go as far as it used to. Nevertheless they have more of it.

Why do prices go up? And who is responsible for keeping prices down? Many blame the butcher, but he is only a link in the chain of distribution. It's hardly fair to vent spite on him. A world situation is responsible.

Let us look at the position in general. At the end of 1972, half the wheat that we used was imported mainly from North America and Australia. Over the previous ten years, supply had been in so much surplus to demand, that in order to protect prices, the American farmer took many millions of acres out of production and left them barren. The term used to describe this was "LAND

BANK". I've personally seen countless acres for which the farmer was paid year after year. There were no crops on it. It was doing nothing. The only stipulation was that the land had to be mown at least once a year to prevent it from going wild.

You may feel why not let prices take their course – and if there's a surplus, why shouldn't everyone benefit? I'm afraid the harsh reality is that it doesn't work that way. If a situation allows prices to become abysmally low, the farmer stops producing that particular product altogether, thus in a few years it disappears and the glut becomes a famine. Let's illustrate this point with eggs again. If prices dipped to the extent that poultry people couldn't make a living, the inevitable consequence would be that they would kill their hens and destroy the battery cages. In a very short time the housewife who has known only a feast of cheap eggs would find that they were almost non-existent, or very expensive.

What I'm saying is that ideally it is necessary to pay a reasonable price to ensure regularity of supplies. Workers in all industries today rightly expect to be well-paid. They accept that the end product which they are manufacturing will consequently be more expensive. It has often puzzled me why they don't relate this to food. Perhaps one of the reasons is that they see such extreme conditions brought about by market fluctuations. A recent example was the storm that arose over the cheap butter that found its way to Russia due to the glut in Common Market countries. This situation sat side by side with the high priced home product. Paradoxes of this nature are not easy to resolve.

Hence, when there is an over-production the question is often raised as to why it isn't given away to a starving country. Again, it's a matter of over-simplification. When there are disasters and catastrophies in any part of the world, it becomes essential to take action along these lines. All that is humanly possible must be done to alleviate suffering caused through floods and other of nature's more fearful manifestations. But, leaving these aside, it does no service to the receiving country to have regular free gifts. Charity doesn't work in the long run. If an emerging nation is starving, hand-outs are not the solution to the basic problem. A country has to learn to stand on its own feet. In a poor country, most people work on the land anyway. By giving them a free supply of commodities they could grow, we are reducing the value of what can be produced on their own native soil. Maybe that's what the cynic meant who first penned the words 'charity begins at home'. It is certainly a complex problem.

To add to the difficulty, we in the western world are eating more than our forefathers. Not only are there greater numbers of us, but also we have 'wealth', created by the printing presses of the Bank of England. If we were paid less, our combined purchasing power would go down with a bump, and there would be more food about, but the situation would be considered socially unacceptable today. Deflation is not much joy to people living on next month's pay cheque. A sharp reverse in prices would put half the nation in economic queer street.

I'm afraid the truth is that most of us are greedy, although we use a word less harsh to describe our motives. It's natural, that as pay packets get fatter and fatter, our bodies tend to do the same, and the desire to acquire articles to make life more enjoyable also increases. Thus the circle is kept turning at a merry pace. And we mustn't forget that food is part of life's roundabout. I'm not defending low wages – just the opposite. For instance, if farm rents had been kept at artificially low levels, the outcome would have been that British farming would not have shown such a dramatic improvement. Likewise, if many farmers' wives hadn't badgered their husbands for what in the past would have been considered scandalous luxuries – washing machines, fridges, colour televisions and the like, the man of the soil would not have been forced out of his previous complacent state of mind.

After ten years of surplus, this year has seen circumstances combine to change the picture. There are two times to sow wheat and other cereals. Autumn and Spring. What is sown in winter, germinates, grows and is followed by a period of dormancy. With the arrival of spring the crop starts growing again, and in the main, winter sown cereals out-yield spring sown cereals. The former can be planted in England up to the end of January, but must always have a period of check before the spring wakes them up again. Nature will hold sway in agriculture, as in all her domain. If one plants winter cereals, they will withstand that season, but if spring cereals are planted in winter, the weather will kill off the crop.

To illustrate this, I'm reminded of a story I heard about Russia. I've never been to that country, but the essential fact is that crops grown there must be covered in snow, which forms a protection. The Russian winter of 1971 was a terrible one from that point of view, but not because of the ferocity of the weather. The trouble was that not enough snow had fallen to form a blanket over the crops and the resultant lack of harvest brought near starvation. The Russians had to buy 11,000,000 tons on the world market, and at the same time, the Chinese harvest hit drought and they had to buy a lot of Australian corn. Bang went the surplus!

Sometimes I think we're getting so advanced that everything can be organised by our technological know-how, but in the end man is puny and his power small compared to the forces around him.

Corn also feeds animals, but, due to climatic circumstances in 1971 the price of this product shot up 30%. The price spiral took its inevitable course. The costs of animal feeding stuffs shot upwards, and fish meal doubled. Following on, beef prices were also affected, as were most meats.

As the standard of living goes up in Britain and the western world, people eat more beef. Consequently, in western Europe the price of meat has gone up by more than 23% in recent times. Demand and supply is still the golden rule of economics. Politicians of various colours have dreamed up answers to overcome this law, but none of them work in practice.

Let me tell you a story. I have a relative, Harry MacPhail, who went to work in Beunos Aires as a railway man. It was in the far off happy days when Britain was a big creditor nation instead of a debtor, and we owned the Argentinian railways. The total picture has changed since those pre-war days and for the better in many respects. Especially as far as living standards in this country go. However, to get back to my story. During those hungry times immediately following the second world war, we had to exchange our railway asset for several shiploads of meat.

To bring the situation up to date, Harry wrote to me only last week, and I was amazed to learn that in the Argentine today, every other week is meatless. This leaves them enough for vital exports. In other words, the Argentine is a dead duck as far as limitless supplies to Britain is concerned.

People think of cattle in two categories, beef breeds and dairy breeds, but the fact is that most beef comes from the dairy herds in western Europe, unlike Texas where they are all beef breeds. The specialist type of animal, such as a Hereford or an Aberdeen Angus is mainly used for providing bulls to mate with dairy cows, mainly Freisian. The older breeds in England are being strongly challenged by European breeds like the Charollais and Simmenthal, which, having only recently retired from pulling carts and ploughs, evidently have well developed hind quarters. This feature commands the best prices in the meat market. Another popular illusion that's fading, is the old idea that nice joints of beef have to be coated with thick fat. Traditional opinion of what is good beef and what is second-class beef has changed somewhat over the last few decades. Nobody wants fat today, but beef can be described as follows:

1. Quick growth.
2. High proportion of lean meat.
3. Most of the flesh on the back end rather than on the front end – the opposite in fact, of Mr. Universe!

Farmers have their fads, too, which bring about conditions of over-production, or under-production. In the dairy business, we get cycles of butter surplus and the type of resultant embarrassment I spoke of earlier. There have been attempts to prevent such phenomena, especially by French farmers who have killed off large numbers of cows, but this accelerated the beef shortage, especially on the continent. We're back again to cause and effect. British beef, being cheaper than anywhere else in Europe, brought to Britain a spate of foreign buyers to buy live cattle to send home. British beef, bought for export, went up by 30% between January '72 and January '73.

Consumer associations and others often hint that the ultimate power is in the hands of the housewife. She has the option to switch from beef to lamb, or any alternative product. But, if the price of our traditional roast continues to go up and up, the hard truth seems to be that such boycotts are neither workable nor effective in the long run.

There is a multitude of pet theories on how to deal with this inflationary situation. Some even suggest that farms should be included in the national freeze, but this is really impractical without import control. Are we prepared to put more of our national resources into farming to provide food at more cost? If we didn't have a stable farm industry, Britain would become more vulnerable to temporary world shortages. Are we to go back to subsidies out of taxation? If we fix prices it means a return to rationing; if we put an end to the free market, where we sell to the highest bidder, we must all be on allocation.

However, as Karl Marx said, "For every action there is a reaction". How long will it be before beef prices tumble down? An impossible dream? Not quite. Perhaps I've made you sit up by stressing that it's our affluence that has brought about shortage and high prices. I painted the picture in stark realistic terms at the time of writing, but there are experts who speak of a slump and falling beef prices. Some go as far as to say that prices could take a downward turn much earlier than most pundits forecast.

Amongst the reasons put forward are the proposed EEC incentives to produce beef from calves, which might otherwise go to veal. At the moment about one third of calves produced in EEC countries end up as this delicacy. There is also a large scale tendency by UK producers to retain more beef heifers for breeding. I also hear that Argentine and Australia are expanding their beef herds to record levels. Then, of course, due to scientific advances, there is a reduction in calf mortality. By the time these words appear in print, I wonder what the price situation will be. Whether it's up or down. I hope that you understand that the reasons are highly complex and not to be laid at the door of any particular sector, political or otherwise.

In closing this chapter, I want to touch briefly on the ethics of exporting cattle, to which many uninformed people object on a variety of counts. Leading the field is the subject of cruelty.

Suffering is deplorable, and I would be the last to defend it, but unless legislation on humanitarian grounds applies throughout the European community, the effect in practice would be almost meaningless. If Britain alone adopted a code, the end result would be that more Irish cattle would go to the continent. This would not eliminate suffering and would reduce any gain in the meat supply situation that I've been talking about.

If I haven't done much for the table, I have tried to give you some food for thought. Commodity prices go in cycles and I've always marvelled that in any free society, if a need is strong enough, somebody comes along with the solution. I've already spoken of the optimists who see signs of falling beef prices. The number of cattle in the world is rising at the rate of ten million in a year and so there is some hope. But to end on a personal note, I have doubts of long term price stability with the common agricultural policy that operates a price intervention system in Europe. Sorry, but I can't see a cheap night out at the steak bar in *my* crystal ball!

9

Knitted Steak or Bean Beef

A FOOD technologist recently made what I thought was a most exaggerated statement. "There's no fundamental difference between these so-called new synthetic foods and our daily bread," he said. Newspaper headlines though seemed to indicate that the food of the future would all be produced in a chemical factory, and that farming would go into a steady decline. Alternatively, it could mean a change in the style of farming, for, relieved by the food chemist in his laboratory of the need to produce, as cheaply as possible, vast crops for staple foods, the farmer could revert to an almost feudal system, designed to supply a wide variety of fresh foods within the locality of the farm.

The city dweller would be encouraged to drive out to the farms in his dearly won leisure time, to help till the land. His reward would be a share in the fresh produce of the farm and a healthy outdoor activity to offset the stupour, boredom and obesity too often associated with highly automated factory or office life. In fact, on my own farm already, flocks of people turn up to pick their own strawberries in the season. Not through futuristic planning though. They enjoy themselves, get the strawberries fresh off the plants a darn sight cheaper than they would pay in the shops, and it is a convenient way for me to market my produce.

Reading further than the headlines, however, one finds that the farmer's days are by no means numbered. On the contrary. There seems every indication that he will need to become more and more mechanised, and more and more specialised to produce the foods demanded by a highly technical intensive industry.

"Synthetic Eating", the headlines shout. "The Big Food Revolution – It's Nearer Than You Think", "Synthetic Turkey with Plastic Wishbone and a Laboratory Apple", "Bogus Beef For Kids", "Courtaulds to Sell Knitted Steak". So the Press perpetrate the idea of food from the laboratory, but all these developments are derived from traditional products of agriculture. My food technologist friend was indeed right. To apply the term synthetic is a complete travesty. The farmer sells his wheat to the miller, the miller processses the wheat to make flour, and supplies this to the baker. The baker then subjects it to a technical, albeit traditional process of dough making, fermentation and baking. Although now, I understand, fermentation has been superseded by high energy mixers in bakeries, which perform the work traditionally done by yeast. Who would have thought, a few years ago, that the dough would have been bashed to hell, instead of being tenderly raised by loving microbes. The whole point is that bread is a wheat product, processed in such a way as to make it reasonably appetising (whatever your views on today's prepacked supermarket loaf).

These new foods, the 'Bogus Beef', the 'Artificial Meat' all turn out to be nothing more than processed versions of soya beans or field beans. The processes are new but the crops are traditional, and the idea of creating new textures from traditional crops is, in a certain way, a bit old hat. Monks in China, 1,500 years ago, were drying and compressing mushrooms which, when added to a stew, tasted very much like lumps of meat. The extraction of protein from leaves is not a new idea either, first being proposed in 1773. The development of interesting textures in proteins from vegetable sources was first patented in 1906 by Dr. Kellogg to whom we are indebted for the cornflake, and the new folk music of the telly commercial. So, even meat substitutes are not new.

What *is* new are the techniques used to make meat substitutes, and I think this is where bread and the new foods do differ fundamentally. That is, the traditional processing of, say cereals, have all been methods which could be reasonably easily carried out in most domestic kitchens, although the modern bread plant looks more like a chemical processing factory.

The new processes, however, are highly technical and can only be carried out in a specially designed plant, with hordes of white coated food technologists pressing buttons and checking quality. Even so, I must agree that the tag 'synthetic', applied to these foods, is very misleading, because for most of us, this invokes a picture of wild-eyed chemists blending foul smelling concoctions in bubbling glass apparatus.

The idea that an appetising human food is being produced from the humble field bean is far more reassuring, particularly for the British farmer. With the pre-eminence of soya beans in this type of development, he must have been wondering whether his American counterpart was going to corner the market and put him out of business, yet despite this reasssurance that farmers' produce will feature large in the age of total technology food, I still feel some qualms, because we are now being offered in the food for animals in this country, protein which has been derived from oil.

The oil companies are very quick to point out that this is not a mystical transmutation of oil into protein, like some process of alchemy, or even a complex biochemical synthesis of proteins, but merely the growth of yeast in a medium containing oil, and it is the oil which supplies the element of carbon essential to all living cells. This is still not very reassuring to the farmer, because yeast from brewing and baking processes, etc., has been consumed for many years. All sorts of beneficial properties have been ascribed to its protein and vitamins. To me it seems easy to visualise food companies eventually using some form of processed yeast, derived from oil, as a basis for a great number of foods, thereby bypassing the farmer again, but here again, my technical friend was able to put this into perspective, and to give a little encouragement to the farmer. There are limitations to the amount of yeast which can be eaten because it contains a lot of nucleo-protein. Too much nucleo-protein – in lay language, too much rich food of the wrong type – can cause gout-like discomforts. Wine fermented by yeast is an example.

In a single cell organism, the nucleus is larger in relation to the whole than in multi-cellular creatures or plants. The nucleus contains nucleic acids, an excess of which is unsuitable for human digestion. This is what causes the trouble.

Apparently this is true of all microbes and there are quite a lot of projects all over the world, in which microbiologists are growing microbes on all sorts of materials, particularly, in recent years, on the waste products of industry and agriculture. Scientists are developing methods of removing these nucleo-proteins so that the protein from the microbes is more wholesome as a human food. In the meantime, we feed these yeasts to our animals who seem not to object, or even if they do, they're not very vocal about it.

In fairness to the oil companies, I understand that they have, for a good many years, fed very privileged and cossetted animals on their yeast proteins, and have observed no untoward reaction in the animals. This does seem to me though, to throw up a big question mark over the field of consumer research amongst animals. How do you ask an animal whether he's enjoying what he's eating?

Production of meat, milk or eggs, in relation to feed costs is the farmers' main criterion. Perhaps they should take a lead from the eight, or is it nine, out of ten cat owners, who give their pussies a selection. Do not imagine that animals have a native instinct for the right diet any more than human beings. If they had, there would be no overweight dogs, or cows dying from bloat through over-indulgence.

Where human beings are concerned, again there has been no adverse reactions in those who have eaten the meat from the animals who have eaten the oil-grown yeast. The technologists think, therefore, that in five or ten years time there will be no reasons why humans should not eat the oil-grown yeast directly – after the nucleo-protein has been removed, of course.

All this research seems rather pointless to me, though, if the fossil fuels from which the oils are derived, are going to dry up before the end of this century. I did learn that if all the available oils for growing yeast were put to this purpose, instead of being used for running cars or central heating, the amount of protein which could be produced would not fill the world's short-fall of protein in the next 20 or 30 years. As it seems most unlikely that people will give up their cars or central heating, I cannot see that we farmers have a lot to fear.

I can, however, see a lot of sense in growing microbes on the waste matter from industries, such as food processing and paper making, along with garbage and sewage of cities. Urban ordure, in fact. This means of obtaining a second crop of useful food after the farmers' produce has achieved its primary function as food for humans or as industrial raw material, must be a good thing to increase the food supply in this country, and hopefully to reduce food imports. I have doubts whether people would find dried microbes a very attractive source of food, but apparently food technology can again supply the answer.

We already eat, in relatively small quantities, a number of these lower forms of life, such as bacteria which are present in milk, cheese, yogurt and the like, fungi which we eat as mushrooms, algae (which incidentally use sunlight as a source of growth energy) which is harvested from certain lakes in Mexico and eaten locally, and yeasts which I have already mentioned. What the food technologists propose, is to take these micro organisms (having presumably removed the nucleo-proteins), and dress them up, so to speak, in a form resembling well-known foods. We're back to synthetic meat again, or what the food people euphemistically call textured protein foods.

As a farmer I don't mind the thought of adapting my animal rearing activities, and I do understand the reasoning which tells us that five times as many people can be fed from one acre of land, growing crops for direct consumption, than the same area growing crops to feed animals. The soya bean people contend that the figure could be as high as 30 times more people, but I somehow feel that they have ignored the need for crop or stock rotation.

No one, least of all me, likes his food to be mucked about with. There's such a built-in resistance to anything which is completely unfamiliar to our normal eating habits that there's little encouragement for the person presented with a bag full of field beans or a tin of dried powdered bacteria. It is an historical fact, however, that under starvation conditions or under siege, for example in the Irish famine, people will readily eat foods which normally they would find repulsive. People have been known to eat rats and mice, and moss and grass. I cannot imagine these conditions occurring in Britain unless some unforseen disaster occurs, and it still seems to be no encouragement to eat field beans or powdered bacteria. The main barrier against their use is that they're unfamiliar and bear no resemblance to existing foods.

I gather the technologists are now spinning, extruding or otherwise manipulating the protein from our beans, and adding appropriate flavours and colours to give the appearance and taste somewhat like meat. In five or ten years time they are likely to be doing the same with proteins from yeasts, bacteria and the like, without those nucleo-proteins. I have seen and tasted textured proteins and I must say they're not at all bad. Perhaps not as memorable as roast rib of beef or chicken supreme, but quite acceptable as run of the mill food, and certainly not objectionable.

It is somewhat chastening to realise that much of the material I feed to my animals could be processed into a perfectly acceptable part of most people's diet. Comparing the amount of protein in the feed with the amount I take out and send to market in the form of meat or milk, brings home the fact that the passage of the food through the animal is somewhat wasteful. However, I cannot imagine that certain items of an animal's diet – grass, for instance – would ever be likely to prove appetising or economic sources of protein for humans, despite the efforts of certain scientists to produce the mechanical cow.

In areas which are difficult to farm with arable crops, grass will, of course, still be utilised by animals producing meat. National Parks, for instance, contain a large acreage of grassland, and sheep always do a better job than lawnmowers. Moreover, in lowland parts of the country, where there is a need for farmers to plant break crops like grass and clover, there will be a need for animals to eat them, so taking their part in achieving the most effective yields from agriculture.

There is certainly a growing concern about the large proportion of the arable crops which we farmers grow, but return for use as animal feed, which could otherwise have been used directly as human food.

If the food technologists achieve their aims, I suppose farmers will come to be regarded by the nutritionists, economists and food processors alike, as biochemical production units, rather than crop producers. Farmers may be expected to produce to a pattern which gives the maximum nutritional yield of food from a farm, regardless of its taste, then the food technologist will take over and sculpt it into breakfast, lunch and high tea. Although I certainly don't relish the prospect, some sort of move towards this situation seems inevitable, even if, as I hope, it will not become the sole raison d'être of farming.

Not so many years back, few people in Britain could have conceived textured vegetable protein chunks as being anything other than a possible means of feeding starving children in underprivileged countries. Yet, British people are now starting to eat these meatless meats in canteens and restaurants, and even some of the packaged foods in the supermarket contain them. The day doesn't seem too far off when British housewives, instead of going to the butchers for a pound of stewing steak, will pop to the supermarket for a pound of 'vegpro' chunks, and new cookery books on textured proteins will be published.

Farmers should be delighted that crops from their own fields are required as raw materials for these new foods, but it seems a mixed blessing, until one thinks of the ultimate horror – synthetic food which has been entirely created in a chemical plant, and has never seen a field.

Doctor Magnus Pike has already given us a glimpse of what is possible, in his book on synthetic food. The farmer's salvation in this matter, as far as I can see, is that the cost of chemical plants to synthesise our food would be so vast, that food would become many times more costly than that produced by Mother Nature in the fields.

To return to the textured proteins being offered today, the designers have obviously used beef, pork and chicken as the blueprints. They should take a lead from the history books, and heed the lessons which other substitute foods should have taught them. Margarine, for instance, which was developed as a cheap replacement or substitute for butter, in the late 19th century. It never became really acceptable until the last few years, when it has been sold as a food in its own right. It seems to me that the sooner food manufacturers establish that textured vegetable proteins are unique foods, worthy of consideration for their own sake, and not as a substitute for our much maligned meat animals, the sooner they are likely to become socially accepted.

The example of margarine also affords us another side light on the effect which convenience foods have on traditional farm produced foods. Another example which foreshadows our farms as biochemical production units – Shades of 1984? The convenience of being able to spread margarine straight from the fridge, on the pappy stuff that passes for bread these days, has a lot to do with the success it has enjoyed over the last few years. Now, would you believe it, the butter manufacturers are suggesting that we milk producers should give our cows modified feeds, which will promote the presence of relatively low melt point fats in the butter, giving butter the same cold spread characteristics as marge.

There are, of course, any number of examples in which the flesh and product of farm animals have been manipulated by breeding and feeding. For instance, the control of the fat level in pigs to produce animals either for ice cream and sausages, or for bacon; rations and hormones for rapid production of plump turkeys and broilers; milk feeds for light coloured veal; vitamins in the feed for brightly coloured egg yolks, and, very recently, flavour injections for tastier boiling fowl.

This degree of manipulation leaves little room for sentiment in animal rearing on farms today. Perhaps a move towards a less intensive reliance of animals in our diet mightn't be a bad thing. After all, farmers are basically trying to trap solar energy into some kind of tangible food to sustain human life. We must all keep our eyes open for any technique which will further the objective, without prejudice to any new method.

10

Population versus Food Resources

DOOM-WATCHING, it seems, has become one of the most popular academic pastimes for the 1970s. Scarcely a week goes by without yet another international conference or university professor or United Nations agency, making more predictions about the population catastrophe which must hit the human race if it continues to proliferate at its current rate.

So often do we hear this view that I suspect that many of us are lulled into a complacent or even a cynical attitude. In fact it has even become a cliché to say that population growth is the most critical problem facing man today. The facts are, however, worth repeating.

The world population in 1650 is estimated at about 500 million. By 1850 it doubled to about 1,000 million, and by about 1928 it doubled again to about 2,000 million. During the next year or two it seems almost certain to double again – to about 4,000 million.

Even without the help of the ubiquitous computer, it is not difficult to realise that if the population doubles many more times, there will not be enough room physically for everyone to lie down at the same time. In fact, the experts say that if we double 16 more times, there will be only 1 square yard of earth for each man, woman and child. Something has clearly got to happen long before this situation is reached, and this is where the experts differ.

When will the catastrophe occur? Which of our natural resources will dry up first and limit our expansion? Which form of pollution will produce the most significant effect in reversing the population trend? Will there be a fearful catastrophe which will reduce the population dramatically, or will there be a steady decline?

All the indications seem to point to a levelling out of birth and death rates some time during the 21st Century. In the meantime, the race is on between the agronomists, the farmers and the food technologists on the one hand to produce enough food, and the educationalists and purveyors of birth control on the other, trying to limit the population growth. At the same time I suppose the 'environmental scientists' must be constantly trying to prevent us poisoning ourselves by over-pollution or by exhausting all our natural resources.

I think that there is so much improvement in agricultural methods all over the world, that some form of pollution will put the brakes on population expansion before food supplies become critically short.

The Club of Rome in a project report entitled 'The Limits to Growth', produced some alarming forecasts about the relationships between the inevitably increasing population and the decline in the supply of arable land, due to the demand of an increasingly urbanised, mechanised and motorised population. In my opinion they have ignored completely the potential of many of the new developments in food production.

First of all, it should be possible to increase the production of all currently available land quite considerably. It's been shown, for instance, that if all the land under cultivation were as productive as the farms of Holland, the world could support 60,000 million people on a typical Dutch diet. If the land were managed as it is in Japan, it could support 90,000 million people on a typical Japanese diet. To grow this quantity of food by conventional methods, however, would require vast amounts of fertilizers, which would probably either exhaust the resources of the minerals required to make these fertilizers, or create an insupportable drain on the fast dwindling energy resources of our planet. So, on the surface, what we gain on the roundabouts we lose on the swings, but it does seem possible that plant geneticists can produce plants which will fix their own nitrogen from the atmosphere, like clover, thus avoiding, say, the necessity of adding nitrates from chemical processing plants. Also, under current day conditions, we would need large amounts of pesticides to stop disease and pests decimating the crops, but here again, plant breeders are getting much better at producing disease-resistant varieties, which reduces the need for sprayed-on pesticides.

Plants like soya beans yield good quantities of protein, which can be consumed directly by humans, but can only be grown at present in certain temperate zones, yet already there is work afoot to propagate new varieties which can grow under other climatic conditions. It is not unreasonable, therefore, to expect that before long we shall be able to grow the crops we want, where we want them, thus minimising the effects of climate.

However, I cannot see the plant geneticists controlling the need for water, and this will probably be the limiting factor.

Irrigation must certainly be improved in many parts of the world in order to achieve significant improvements in crop yields, but although the cost of such huge irrigation schemes is enormous, what is it compared with the need for food?

This subject of world irrigation prompts me to raise what I consider to be a ludicrous situation in England. Surely, in a climate such as ours, there should never be a danger of water shortage. Countries with a far lower average rainfall than ours have systems which practically guarantee an adequate supply, in far longer dry periods than we have. Long term planning is the answer, not panic measures when the reservoirs are at danger level.

Flooding good farm land all over the country is not the solution, but a short sighted substitute. It's bad enough to take steep sided valleys in marginally profitable areas, but to store water at a depth of 30 feet can surely only be justified on strategic grounds in case of conventional attack.

De-salination plants, which, I am told, provide water at three times the cost of reserved water at the reservoir (Note, not delivered via massive pipes and pumping equipment), must become more significant. These are just the job in places like Gibraltar, where I visited such a plant, and surely could be sited around our own coasts adjacent to existing industrial complexes.

Perhaps if water was dearer, waste would be less excessive, or have the experts considered methods of recirculating the waste, particularly from industry and from the thousands of automatic car washes. Not only in dry spells, either. Stock the reservoirs to the brim and keep them filled. 'Waste not want not' is more than apt, when water is concerned.

Back to the food problem, I cannot really see the possibility of supporting 90,000 million people – or even 60,000 million people, but it is obvious that agricultural production could be pushed far, far higher on average than it is currently, to adequately feed a very much larger population.

Hopefully, then, we would expect the efforts of birth control agencies, or even the social and psychological pressures created by over-dense population, to curb the birth rate sufficiently to stem the expansion of the world population before catastrophic disaster overtakes us.

Here is another example which illustrates that man has great potential for producing very much more food from the same, or even less land area.

The present world average agricultural productivity requires about 1 acre of arable land to provide the food necessary to sustain one person, but, based on the Dutch and Japanese experience, it would appear that about one-sixteenth of an acre is sufficient to produce food for one person. Yet even if agricultural efficiency was improved, but land area still proved to be the limiting factor, we have the opportunity of growing the micro-organisms discussed in the chapter on New Foods, in tanks by processes not dissimilar from the fermentation of beer.

The technologist can produce not only large quantities of yeast, but also, if necessary, bacteria or fungi – which are rather different, I gather, from the mushrooms that grow in the field – and even algae. The advantage of this rather unusual approach to producing nutritious foods is that these micro-organisms can use all manner of waste materials as the food which they require for their own growth. Whereas in beer production, the barley sugar provides the food and energy for the yeast to grow and produce alcohol, in these new forms of food production the yeast, or whatever, uses some of the waste products, or surplus materials, from industrial processes – indeed even from city sewage. For instance, after wood has been processed for paper-making, there is a lot of sugar-type material, the waste liquor, in which yeast can be successfully grown.

The algae which I mentioned are micro-organisms which require sunlight, like plants, to provide the energy for the growth process. This means that algae can be produced without burning up so much of man's not inexhaustible supply of fuels. Yeast and bacteria growth requires larger inputs of energy and minerals.

A well-publicised example of the new tank grown foods is the production of protein from oil. What is not often appreciated is that it is the yeast which uses oil as part of its food source in a typical fermentation process. The yeast is then harvested by the processor; freed from oil, and one is left with a dried yeast which contains 50% of its own weight as nutritious protein.

Quite clearly the enormous amount of food which could be produced from a very small area required for the tanks and equipment of a processing plant would make traditional arable farming look very old-hat.

Professor Arnold Bender of Queen Elizabeth College, London, has even gone as far as to suggest that with this type of set-up we require only one square yard of area to provide the food needs of each man, woman and child on this planet.

The thought that it will be possible to provide the food needs to support almost limitless population growth with adequate nutrition – although perhaps a fine thing – fills me with the utmost sympathy for the future generations who are bound to be subjected to cruel over-crowding, if the proliferation is allowed to continue.

I discussed these developments with Brian Smith, a consultant food technologist I know, who has had a lot of experience of these new protein foods. He pointed out that there are severe limitations even for these uni-cellular foods made by fermentation, not least of which is the problem of converting them from the amorphous and unattractive form in which they are taken from the growth tanks, into an appealing and appetising food for humans. There are also within these micro-organisms, components which are not suitable for the human digestion if eaten in large quantities. Techniques for overcoming these problems have been discussed in my chapter on processed foods.

No, I don't think they are the universal panacea, but they may help to release more of our arable crops for direct consumption by man rather than consumption by his domesticated animals. If the animals eat more of these single-cell protein foods, man will get far more efficient usage from his field crops by consuming them directly himself. At least five times as many people can get their protein needs by eating agricultural crops directly, rather than via the domestic animal and then consuming the meat or milk produced. This is one very good reason for believing that traditional agriculture can still remain as our principal food source, but it's certainly not the only one.

There is the tremendous problem of what to do with the vastly increasing numbers of surplus people in countries with mechanised farming. Clearly not everyone can be used in all the ancillary activities of mankind which have grown up in so-called civilised society. By ancillary activity, I mean anything which is not directly concerned with either feeding us or protecting us from the elements, or raising our children; in short, survival in the most basic forms.

In a highly industrialised country such as Britain, it is frequently not realised how very much of men's time in aggregate is concerned solely with the production of food. In the world as a whole, two thirds of the working population devote their lives to feeding themselves and one another. This means that the ancillary activities to which I referred, are on average the product of one man in three.

In countries which have mechanised food production, one has a rather peculiar situation. In the U.S. for example – one of the world's biggest producers of food – 150 years ago one farm worker fed four other people; one man produced food for five. At the beginning of this century in 1900, the work of this man fed a total of eight; 50 years later one man fed a total of 17 – but between 1950 and 1969 the number increased threefold, so that one man on a farm in the U.S. now feeds a total of 48 people.

This enormous increase in production is due to the combined efforts of animal and plant geneticists, plant and animal nutritionists, the use of irrigation, fertilizers and pesticides – but mainly because of the mechanisation of farming. Now, however, they have arrived at the deplorable situation in which mechanisation has reached such a stage that farming, which is a process after all intended to trap the free energy of sunshine for the benefit of man, costs as much in energy input in the form of fertilizers, chemicals, tractor fuel as is obtained in the form of calories from the food produced.

On this basis the peasant farmers in Nepal – who comprise 92% of the population – with no other aid than the water buffalo, are far more efficient than the 'soil technocrats' of America.

I believe, therefore, in the light of these examples, that man will be driven to reach a compromise between these two extremes, because clearly we cannot continue to use our exhaustible supplies of energy and materials in the production of food at the rates they are being used in the States.

In Britain, for example, we will need to have more of our men working on the land than there are at present, even if only to provide useful occupation for the many made superfluous by an automated society, if they can get there without too many motorways!

There is a Chinese proverb which reads 'If you give a man a fish, you feed him for a day; if you give him several fish, you feed his family for a day. If you give him a fishing line, you feed his family for life'. This sort of philosophy could well be applied more widely, and in the light of it, it should be possible never to fear starvation whatever the world population becomes. If we apply these principles, the limitations to world population should lie very much in areas other than food supplies.

The current rough estimates are that about half of the world's population are badly fed and about 10% are really starving. About one third of the world's children are malnourished. One of the great hopes of the last few years to alleviate this situation, has been the 'green revolution'. New strains of rice, wheat and maize have been developed which give far bigger yields than traditional varieties, when they're appropriately cossetted with fertilizers and pesticides. However, these new 'miracle seeds' have not fulfilled their predicted promise because the societies in the countries where they are needed have had insufficient capital to buy the fertilizers, or to provide the irrigation, or to buy the pesticides, which seem to be required in much larger quantities to reap the benefits of the new varieties.

The result is that rich farmers get richer, but the smaller rural communities and the subsistance farmers get no benefit. In fact, the rich farmers sometimes buy out the small farmers, increasing the numbers of landless peasants without any job of work. I suppose their inevitable fate is to drift to the cities where they join all the other miserable, underfed and unemployed families.

Quite clearly the financial aid programmes of World Health Organisations and world governments should be aimed at providing adequate water and adequate fertilizers, to enable these high-yield crops to be produced at *all* levels of agriculture.

Are these ultra-high yielding strains and varieties too good to be true, I wonder? For instance, is the nutritional value of foods made from these 'miracle crops' as high as that from traditional cereals? I discussed this with Dr. Gordon Booth, a well-known nutritionist, who also has a wealth of experience in all the techniques of food processing. He assured me that with very minor exceptions, the nutritional value was as high, and in protein terms somewhat higher, but he pointed out the problems in the acceptability of some foods made from these high-yielding crops. He quoted the example of tortillas made from high-yield maize, which tend to be a different shade from those made with normal maize, and consequently have run into considerable consumer resistance in South America.

This type of reaction is typical of course, and one which is familiar even to Britain. Here we seem to prefer brown shelled eggs to white, and white bread, preferred by most, to brown bread; custard must always be yellow; reason and logic are by no means necessarily involved in the choice of food. How frequently we hear the old adage 'A little of what you fancy does you good'. It is certainly not always true. You know, man has little instinct for choosing on the basis of how much good it will do him.

Consumer preferences are strange. People tend to think that what is familiar is best. For instance, some always ask for King Edward potatoes, or Bramleys Seedling apples, not because they can tell one chip from another, or the difference between apple tarts, but because they think they are buying better produce.

This is understandable when, according to the telly adverts, some of our brethren are unable to distinguish margarine from butter. Perhaps they are lucky though!

When I sat down to the supper table after the evening milking, there was nothing I enjoyed more than a plate of fresh bread spread with thick layers of butter. It is difficult, though, to retain one's enthusiasm when faced with the link between cholesterol and heart disease – and eating animal fats which are reputedly high in saturated fats. Butter is apparently one of the worst offenders. Dr. Booth told me of two very interesting experiments which seem to provide unshakeable evidence which must give us cause for concern.

In the first experiment, carried out in Finland, 200 permanent patients in one hospital were served for six years on a diet largely free from animal fats. In the second hospital, another 200 patients had a diet which contained all the usual dairy products like butter and cream and other animal fats. They recorded the number of patients who died during this six year period from coronary thrombosis. The hospital where animal fats were consumed, had a significantly higher death rate than the other. The experimenters then reversed the diets so that the 200 originally on animal fat-free diets started to consume butter, cream etc. They continued this trend for another six years, and lo and behold – the incidence of coronary thrombosis also reversed. This painstaking experiment appears to be a fairly strong indictment of animal fats, particularly in dairy products.

I gather that the experiment further reinforced food laws in Finland, which already places restrictions on the sale of certain dairy products.

The second investigation which perhaps will serve as a warning to us all, involves the tracing of 200 pairs of brothers born in Ireland, one of whom had emigrated to the Boston area of the U.S.A. and one of whom had remained in his native Ireland. That, for a start was a pretty remarkable piece of detective work! Then the investigators measured the habits and diet and life styles of each of the brothers. The main differences found were that the emigrants had a much higher calorie intake and a much higher fat intake than the stay-at-home brothers. Also, the Boston brothers walked very little in comparison, and tended to drive almost everywhere. Fairly predictably, the Boston brothers tended to be heavier and, of course, the final conclusion the investigators discovered was a far higher incidence of coronary thrombosis in the Boston brothers.

It appears obvious that the message comes over loud and clear – we should alter our eating habits to consume less meat and dairy products and rely more on arable products. This further reinforces the ecological argument that we should move to a style of agriculture which produces arable crops for direct human consumption.

To get back to this link between our food supplies and the population problem; few of us feel that its effect will be noticeable for many years yet – and certainly farming attitudes in this country are influenced more by what is commercial, profitable, or even traditional, rather than what is a social need. But British farmers produce only half the food for our consumption; the rest is imported. This is fine as long as the exporting countries can afford to send the food to us, and we have the credits to pay for it. But what happens when those countries suddenly find they need all the food for their own people, or perhaps their less fortunate neighbouring countries? I suppose that if we are prepared to pay a high enough price we can fly in the face of the social needs of others, and obtain the food by the sheer force of our buying power, but I cannot see that situation being sustained for long.

The relationship between race and level of intelligence has been hotly debated recently. Whatever your views may be on the relationship between I.Q. and genetics, there is no doubt that an under-fed or under-nourished man can bring neither the same intellect nor physical ability to bear, to solve his problems and to improve the miserable situation in which he finds himself.

Dr. Booth reports that evidence has recently been published to the effect that mental and physical capacity is closely related to the amount and quality of diet. When this is fully appreciated by the leaders of countries with a large proportion of under-fed inhabitants, they must surely agitate strongly for a more equable distribution of world food resources, and also scientific and technical resources which are available to advise on the methods of improving agricultural yields.

This is their most valuable means of removing the inertia which seems to bedevil any attempts these countries make to pull themselves up by their own bootstraps.

11

Health Foods - Fact or Fancy?

'HEALTH FOOD' has always struck me as being a most unnecessary phrase. It seems to imply that any food which can't be so described is somehow unhealthy. At best, it's a phrase perpetrated by those who sincerely believe that such food does have some mystical health giving quality, and at worst, it seems to be an ill concealed excuse for charging inflated prices for perfectly normal food. I will grant, though, that some of the food described as health food is probably grown without resort to chemical pesticides and so-called artificial fertilizers.

One factor which seems to be missed completely by the people who cling to the organic food ethic, is that lime is an essential ingredient in the soil for growing the vast majority of crops and plants. Now lime is an inorganic material, it is added as an inorganic material and it is usually an essential part of manuring the land. Therefore, when the health food people talk about using only organic manures, they are completely neglecting one of the most essential.

It must be obvious to everyone who has ever done any gardening, that to get a good yield from one's soil it must be well balanced in terms of the various components, which include such things as humus. Certainly, every farmer must understand this. To listen to what some of the health food people say, one would begin to think that some of the plants actually gobbled up any artificial fertilizer added to the soil, which in turn would then be eaten by humans when that particular crop is consumed. However, the fact is that the fertilizers are often converted by the bacteria present in the soil to make them available to the plant, frequently in an organic form,

The kids. Patrick, Anna, William, Tommy, Ellen-Mary, Amelia.

BOTH SIDES OF LIFE

Above : Piping a spring to give the cows a drink.

Top Right : Compering a fashion show at Cheltenham.

Bottom Right : A nice job to have. One of the other two is John Jennings, M.P. for Burton-on-Trent.

BARGATES

What's My Line — Isobel Barnett, self, Sara Leighton, Gilbert Harding. BBC copyright photograph.

After a charity performance of Ask Me Another, at Douglas I.O.M. Franklin Engelmann, next to me, was chairman, and fellow panellists were Olive Stephens (centre) and Dr. Reginald Webster (2nd from right).

Doing a commercial for the Milk Marketing Board.

Blodwyn shows what she thinks of barbed wire.

Off to America with the pig and poultry experts.

Just taking it easy — down to earth.

Not taking it easy! Helping the Fire Service to control a fire on Tom Field's farm.

Our House.

View from kitchen window.

even though added to the soil in inorganic form. Neither is it true that all organic material in the soil is good and useful for the plant. The molecules, I am told, in much organic material are too large to be absorbed directly by the plant, until they have been broken down sufficiently to pass through the root hairs. This is achieved, again, largely by the bacteria in the soil. This screening of material by the plant means that either way, organic or inorganic, the end result is likely to be about the same, unless, of course, the plant lacks any particular nutrient. I can say from personal experience that it's very much easier to arrange and organise a balanced soil with artificial fertilizers than by using farmyard manure.

I can certainly understand, and even share, people's concern over the relationship between the food they eat and their health. After all, man is what he eats. Also, medical scientists and nutritionists have clearly demonstrated the link between our food and certain diseases. Apart from the obvious supply of energy and body building materials from the fats, carbohydrates and proteins, our food also supplies the vitamins and trace nutrients which seem to be necessary to prevent us contracting diseases such as scurvy. People should be taught something about nutrition so that they can select their food more judiciously. This is probably more important than ever, now that food tends to be selected more on the basis of convenience than by what happens to be in season, or what is traditional. This is, perhaps, the biggest failing of health food fadism. In their excessive concern for the very small and sometimes spurious differences which may exist between 'their type of food' and every one else's, they seem to miss the major essentials of eating for health.

Take sugar for example. It is apparently all right, the health food fadist will tell you, to eat so called raw demerara sugar or Barbados sugar, but it is not O.K. to eat raw white sugar. In point of fact the differences between the two in dietary terms is almost negligible. What is significant, is that every man, woman and child in this country, eats over 100 lbs. a year of sugar, whether it be white or brown, and as people like Prof. Yudkin and other nutritional scientists have pointed out, we are much more likely to suffer health problems through eating this excessive amount of sucrose, than we are likely to benefit by consuming the minerals and other gunge which constitute less than 1% of brown sugar.

I have no animosity towards health food addicts, in fact, as the cannibal said "some of my best friends were vegetarians" but I think they have perhaps inherited the distrust of foods which grew up when people in the middle ages were subjected to all sorts of horrifying sophistication and adulteration of their food. Chalk was added to flour to make it look white. Lead compounds were used to colour food. Sand was used in pepper to make it go further. Sugar and water were added to ale to make it go further. They were able to check on this last one, however, by sending out gents in leather breeches to buy their samples of ale at each ale house. They would then pour it over a wooden stool and sit in the puddle formed. After half an hour they stood up, and if the stool stuck to their breeches, the publican was prosecuted. Apart from this, there was only a slight risk of discovery because analytical methods were so little developed during the middle ages. Sometimes, poisonous materials found their way into foods and drink by accident. For example, cider, when brewed in enamel vessels, dissolved the lead from the enamel and caused many cases of lead poisoning.

However, once the techniques of checking and control were developed by chemists, one had the beginning of pure food laws, which now quite strictly control what can and cannot be added to foods and what must and must not be present. So, although the idea of health food possibly grew up by reason of fears handed down from father to son, these are no longer nearly as appropriate as they once were.

One of our basic foods, bread, is a good example of the way in which fashions in food are sometimes in conflict with nutritional knowledge. Back in the middle ages again, flour was sifted on silk and the finer and whiter flour was often used by the more affluent. The rougher tailings, including the bran and so on, were consumed by the poorer folk. In those days there was no fad about wholemeal bread or whole grain bread, it was simply that those who could afford it, went for the whiter bread which was more agreeable to their palate. To-day, although most people will agree that there is some benefit to be gained from the bran present in wholemeal bread, but absent in white bread, lip service is paid to this knowledge and the consumption of white bread is very much greater than that of wholemeal bread. Health foods shops cash in on this by charging more for the wholemeal bread *they* sell, than the white bread the supermarkets sell, and they will charge even more if they claim that the wheat from which the wholemeal flour is made was organically grown.

This I can understand, because farming the organic way is almost inevitably more costly, I don't think the consumer benefits because there seems to be no difference between the nutritional value of organically grown wheat and its conventionally grown counterpart.

An aspect of the health food creed which completely mystifies me is the arrogant assumption that all things natural must be good and safe and healthy, and that all things chemical must be harmful. A fact that is ignored, is that there's a whole string of examples of natural foods which contain some toxic hazard. There's the solanine in potatoes developed during bad storage, which has been linked to spina bifida. There's the oxalic acid in rhubarb and other vegetables. There's the cyanide in damsons. Now most of us are able to adopt a fairly reasonable attitude to these natural toxins even if we know they are there. We tell ourselves that in the quantities we are likely to consume, they're not likely to do us much harm, but when it comes to trace food additives such as food preservatives or emulsifiers, we become irrational, despite the reassurance given us that extensive testing has shown that the only additives permitted to be allowed into our foods are those which have been demonstrated to be safe at relatively large doses. We despise the food chemists who have added them. Consumer watch dogs accuse the food processors of slow poisoning of the population, but scant regard is paid to the benefits which these additives clearly provide in our processed foods.

Don't imagine that I am particularly fond of adding chemicals myself, but when one realises the benefits which have been achieved, not only by food additives, but by pesticides and drugs, mankind really owes a debt of gratitude to the chemist.

If you still feel there are slight risks involved, these are surely outweighed by the elimination of a much more serious risk of food poisoning or disease. Take for example bacon and ham. They now tell us that the nitrates and nitrites, which have been added to curing solutions in this country for centuries, may give rise to traces of cancerous compounds. It seems much more likely, however, that many more people will die from botulism (or food poisoning from eating contaminated meat), due to infection by microbes which are less readily held in check when the nitrates are absent. An even more obvious example is DDT. Malaria has been eliminated from vast areas of the world, and goodness knows how many millions of tons of crops have been preserved for consumption by humans rather than by pests, by the use of DDT. Yet, despite the fact that no death or case of cancer has ever been attributed to DDT, even amongst many thousands of people who are employed to spray the pesticide, public opinion was whipped up to demanding a government ban on its use. This was based on what I now gather is the rather flimsy evidence, that DDT killed some birds and made their egg shells thinner, and that it was even found in some penguins. If you think that my attitude seems somewhat complacent, you should remember that before DDT arrived, three quarters of a million people a year in India alone, died from malaria. Many more eagles and other wild birds are killed by hunters with guns than by pesticide residues, and if DDT can help to ensure that those who have survived malaria also survive starvation, then I'm all for it.

There are, of course, some unmitigated disasters such as thalidomide, where the responsibility for human suffering is surely on the shoulders of the scientists. Yet, when you stop to compare the suffering, misery and even death, in the rough and tumble of man against nature in her alarming extremes of earthquake, frost and fire, and in the trial and error which must have led up to choosing the right foods or the right herbal remedies or even the right witch doctor, I think the occasional errors of the scientists take on their true perspective. With the present world population, you just cannot hold up scientific advancement intended for the benefit of mankind, albeit with a profit motive, without committing many millions to some form of suffering, either through disease or starvation.

I mentioned the profit motive because even farmers tend to grow the crops and graze the animals which yield the best profits. I can see a real danger that if conservationists, ecologists, health food faddists and the like are too vociferous and too successful in their agitation, much of the research primarily done for profit in the fields of pesticides, drugs and food preservatives, will be ended by the companies involved, who will search for less vulnerable areas for their profitable business.

This would undoubtedly slow down the rate of innovation and improvement, which I think would be a pity for mankind. Every time a health food champion rubs his hands with glee about a new ban on a food additive, he should perhaps spare a thought for his fellow men who might, in fact, be worse off without it. Take, for example, the case of cyclamates, the artificial sweetener, which must have made foods very much more palatable for many thousands of diabetics, and much more tolerable for countless obesity sufferers. Although no harm to humans has ever been shown through consumption of cyclamates, the United States, followed by the rest of the world, banned cyclamates in foods on the basis of one very dubious study carried out on animals. Now, I understand, the other major artificial sweetener, saccharine, is itself coming under the same sort of suspicion. If saccharine *is* bad, consider the suffering of diabetics, and isn't a ban likely to increase the amount of sugar consumed by people, and isn't this much more likely to produce obesity, heart disease and other complaints?

Whilst on the subject of sweeteners, consider the old chestnut type debate on the relative merits of brown and white sugar. They are chemically identical, both being sucrose, although whereas white sugar is virtually 100% sucrose, there is more gunge in the brown variety. This gunge is mainly minerals, a little protein and fibre from the sugar cane, plus assorted insects and rubbish which happened to be around at the time. Brown sugar does contain a chromium trace which may be some good, but surely the important factor is to limit the total intake of sugar, whether it be white or brown.

Too many health food people think only of themselves when they're pontificating on health foods. In the context of the 1970s surely they must acknowledge we need food additives. When we have the world population under control, and down to something like a comfortable level, then there will be more room for the agricultural principles and food mores of the health food clan.

One thing I have noticed in talking about health foods is that there are definite double standards. Amongst health food advocates there are recognisable groups such as vegetarians, vegans, macrobioticists, organicists, herbalists and soil associates who have reasoned with mainly acceptable arguments for their particular beliefs and attitudes. I do not, however, agree with all their ideas. On the other hand, there are the health food manufacturers and retailers, for whom the definition 'health food' appears to be anything which is sold in a health food shop. These products range from the frankly crank things like biochemic tissue salts, pollen extracts and the vastly expensive universal panacea elixirs (of life?) to more normal foods such as cornflakes, wheat flour and salt, usually sold at inflated prices.

Why we should have to pay so much more for sea salt than for ordinary salt I cannot imagine, especially if the salt contains a concentrated form of the trace metals with which our effluents are apparently polluting the seas.

In one of the tracts issued by the group supporting the "organic" foods, one will frequently find passages berating the conventional food manufacturing industry for rendering our food worthless by such processes as freeze drying. Yet one health food supplier, who professes to purvey only organically grown foods, offers freeze-dried yogurt, and even yogurt tablets. To those habitual eaters of fresh yogurt, it must be a comforting thought that they will have a longer life due to their guts being colonised by friendly bacteria. Bacteria, the multiplication of which, according to one of their pamphlets, "inhibit the activity of micro organisms responsible for the slow poisoning of the blood stream and vital organs of the body".

How these friendly creatures survive the drying and tabletting process is not indicated.

Isn't it true that today we live, and expect to live, a lot longer than our predecessors only 50 or 100 years ago? I know a lot of this is due to medical science, but it also coincides with the upsurge in processed, convenience foods, pure food laws and an increase in the watchfulness of public health authorities on the standards of food and the water we drink. No, I really cannot believe that the consumption of health foods is going to make us significantly healthier than if we consume the usual range of processed foods, backed up by the fresh fruits, and vegetables and meat which most people buy from their local greengrocer and butcher.

One thing that makes me wonder about the 'ban pesticides' movement is how far the supporters of the movement are prepared to share their food with the pests. As one critic pointed out, you can ban pesticides but you can't ban pests! It's all very well to have a few hundred farmers, in isolated pockets all over the country, raising crops and animals without the benefits of pesticides and chemical fertilizers, but what would happen if all farmers went over to this style of agriculture? If we allowed the pests to share our crops, dispensed with our guns and traps to allow the rabbits and the pigeons to eat their fill, and invited all the wild birds to come and take their share, I can hardly see much being left for human consumption. Also, in the 'free society for pests' there would surely be amongst the birds, rodents, insects and micro-organisms which were permitted to flourish, some which would inevitably constitute a serious hazard or threat to man himself, apart from the problem of not leaving him enough food.

This is just another example of why I think health food people tend to think only of themselves. The beliefs of most of their groups can be applied in our society, only to a small minority. Any reasonable person must realize that we cannot again allow the conditions to arise, which allowed the scourges of past centuries such as cholera, bubonic plague, potato blight and so on to sweep across the country.

There are many other anomalies perpetrated by the exponents of organic agriculture. They recommend soot and bonfire ash as fertilizers for crops, yet these materials are loaded with cancer forming materials from the smoke of the fire, which might find their way into the 'organic crops', whereas the ammonium sulphate which I add to my land is certainly not carconogenic.

Take again the instance of the cyclamate ban. This ban was based on the premise that cyclamates were essentially cancer forming materials. On the same basis, many of our common foodstuffs would be banned because they contain naturally occurring cancer forming compounds, identical to those found in cigarette and bonfire smoke. These would include cabbage, lettuce, leeks, even our good old cup of tea. In addition to the potatoes, rhubarb and damsons, which I've already mentioned, we'd have to ban onions because they contain some anti-vitamin compound which could cause anaemia in excess, strawberries because they contain coumarin which is a flavouring material recently pronounced as unsafe by food toxicologists, and even milk, because many people after childhood show a severe intolerance to its lactose sugar content which causes severe digestive upsets.

If all the above was taken seriously, I might as well pack up until people realised that there was nothing left to eat.

Of course, all of these are natural products, so presumably it doesn't matter. These are the double standards with which the food purist must come to terms. I, for one, consider he should accept that the food we get today is of a pretty high standard, and that any chemicals which are used in association with either the growing or the processing of that food have been pretty thoroughly screened by experts to ensure a good standard of safety. In fact, a standard of safety which in some cases is much higher than that which we enjoy through consuming the 'natural' product.

Surely manufacturers of health food products, and health food shops themselves, are bound to be forced into rather two faced attitudes. On the one hand, there are the inevitably higher prices forced on them due to the growers of their products sharing their crops with the pests, and on the other, their processed products are processed in the same ways as the conventional processing industry, and must consequently share the same shortcomings for which they criticise the conventional industry.

The most significant shortcoming of health foods to me, is the lack of outright proof and hence the lack of outright claims that 'health' foods are specifically better for one than normal foods, or organically grown foods are better than foods grown by my own system. Purveyors of health food are aware of the penalties of misleading claims, and consequently their advertising, or articles supporting health foods, are usually full of vague generalisations and implied rather than stated benefits. This leads me to believe that there is a basic uncertainty about the beliefs of health food faddists and this uncertainty reveals itself as a double standard when companies try to cash in on the fad.

Perhaps I am wrong. Perhaps having written this, someone will come along and give me the proof which I have been unable to find. Perhaps if this happens I might even turn over to farming the 'organic' way. But I've always thought of food as organic anyway.

12

Farmyard Morals

THERE'S nothing like life on a farm to teach kids about sex. After all, cows are big animals and when the time comes for bovine copulation it's not likely to be missed by any young enquiring minds, who happen, by chance, to be near the scene at the time. In any case, neither my wife nor I would have any wish to hide the facts of nature from our six children. Country people readily accept birth and death as the fundamental order of things. A plant grows, it flowers, it seeds, and then wants to die.

The basis of life can be summed up as desire to perpetuate the species, and this is true throughout the whole of nature. Humans are animals, but playing, so to speak in a league of their own.

It's many years since Darwin shook Victorian society out of its cosy attitude by proving almost beyond reasonable doubt, our common descent with monkeys from fundamental life forms. What have we in common with other animals? We have similar bodies and we learn from their behaviour. Domestic animals illustrate in their own life-style the human pattern of frustrations and neurosis. For example, take a herd of cows. Their pattern is very akin to that of our own society. You have the boss cow, who is always the first to trough and pushes her way forward. Then come the untouchables, who are fortunately few. They're always being pushed around. So you see that animals, too, have a top strata and a bottom strata, with a great deal in-between. Perhaps the floating population is best described as proletarians, uncertain of their place in the pecking order.

The comparison with human society goes even deeper. In the bottom section there are the old-aged pensioner type of cows; the sick who are inflicted by disease or injury; and, of course, those without hope, the equivalent of the drop-out in social terms.

Fortunately, most humans have more compassion for their lesser brethren than do animals, who are not concerned with finer feelings. In the wild, if a wolf is injured and not able to participate in the action of the pack, the instinct of the others is to destroy the weak link, thus eliminating a liability to the community. There have, of course, been political leaders whose ideas were not far short of this cruel philosophy.

Throughout my working life I have observed cows at close quarters. Their psychology is more than interesting. The top strata and the bottom strata know where they are, but those in the middle seem to develop anxieties and fears about maintaining their place in the order of things. It's the old story: there are two comfortable places in which to be in this world, namely at the top, or the very bottom. At one end of the scale you can't rise any higher. At the other end, you can't fall any lower. Either position can be comfortable if the mind of the individual is attuned to his circumstances. No-one, however, seems to be happy between the two extremes. Although the main discourse of this chapter is concerned with morals, a bit of background to the situation doesn't come amiss and I want to pursue it awhile before returning to the main theme.

One thing I notice about industry and commerce is that those in the middle executive grades always seem to be looking over their shoulder to see which of their contemporaries is whetting the knife (before his own is sharp enough). One can only feel sorry for such people, but why don't they develop a sense of perspective and basic philosophy, taking a lesson from the mistakes of dumb brutes?

Back to animal farm. Hens and birds have a latent desire to become cannibals. Let's hope we humans don't copy this tendency! Yet, I recently read with horror about a man driven to desperate extremes who ate a dead companion to avoid starvation. Anyone who has kept a deep litter of hens will know that any weak fowl is liable to be gobbled up by his fellows. In the broad sense it is the natural law of the survival of the fittest in action again; but humans are less inclined, thank God, to gobble up the old and the sick. They do try to care in one way or another.

This brings me back to the subject of sex. Although the secrets are laid bare on a farm, they don't have as much advanced teaching value to children as they did twenty years ago. The explanation is simple. Today, all kids understand the mechanism of sex. It's taught to quite young children. This is due to the type of education in schools and a more open attitude to the subject by all the media. Nevertheless, an observant child on a farm can still note the psychological difference between human and animal behaviour. Animals experience desires as strongly as man, but only at certain times, and then for the purpose of procreation of the next generation. Their sexual drive seems to evaporate at other periods, unlike the lusts of man. Nevertheless, anyone who doubts the frenzy of animal passion at the right time would perhaps be surprised to see cows in season turn into Grand National hopes – leaping walls, fences and other obstacles. And the noise they make! Cows really are quiet creatures at other times. If, on a summer night, you hear loud bellowing from a field of cows you'll know it's one of three things: they're in season; in distress; or have broken out of the field.

Deer are a more exotic and even noisier example. Their mating ritual – with complete disregard to my interests, or any other farmers' – consists of trampling down enormous circles of corn in their phrenetic desire to copulate. As I see it, all the sensations that animals experience are limited to self-preservation, sex, food and warmth. But to give the animals their due, it's not sex for sex sake. In this respect they are entirely unhuman.

Children who are brought up on a farm in natural surroundings see the physical aspect of sex uncomplicated by emotions. Farm animals don't have to be in love with one another, or have an emotional relationship in order to perform the sexual act. It's odd to think that the nearest creature to man, whose emotions must be involved is to be found in the bird kingdom. Swans, pheasants, and many other birds, restrict themselves to one partner for the season, if not for the rest of their lives. A lot of wild animals are the same.

As in all other facets of human life, sex has its fashion cycles. We're told today that marriage is on the rocks; gonorrhea should be a socially acceptable disease; pre-marital experience is an essential part of further education; and extra-marital affairs are permissible, even said to strengthen marriage.

We are bombarded with outrageous ideas from every quarter, including the media and commercial advertisers. It seems that sex is the overt factor in promoting products: the chap who drinks Bloggs' beer, or smokes X cigarettes, gets the best girl, as his primary reward. Satisfaction with the product is a secondary consideration. The thought occurs that society is maybe rushing headlong, like Gaderene swine, into a behaviour pattern that would shame the lesser animals and insult their instinctive intelligence.

Reflecting on these disturbing thoughts, I, as a parent am often searching for a slick formula to serve as guide lines for my own brood. Perhaps it's easier to write down my observations, and this book gives me the opportunity.

We've always maintained in our household that there should be no taboos concerning the body or how it works. As our children have grown I have made sure, as my wife has, that our instructions and advice have grown with them. We both think that healthy knowledge is an essential part of child education, unlike Victorian parents who would never mention the 'filthy' subject. When a child of ours asks a question, we answer it factually without adding complications. It's a parent's duty. Sex education in school is fine, but limited.

Every mum and dad is still faced with individual problems. Knowing about the mechanics of the body is one aspect, but there's no escaping the necessity of fully preparing a child for the emotional jungle that he or she must surely encounter from early teens. Most parents duck out and adopt a laisser-faire attitude, but it's really a rejection of their responsibilities to the child. Unfortunately, the corny old phrase 'don't bring trouble' is still bandied around as a piece of classic advice and the subject promptly dropped.

We have found that the way to approach the emotional problem, is to draw on our own experiences and explain to the children how easily emotion can take over from reason and knowledge. Emotion can supersede reason – and therein lies the deadly danger. Teach the child not only to *know*, but also to exercise a degree of self-control and discrimination. It's not a matter of stemming feelings, but of selecting the right person with whom to share them. It's not much good telling the child not to do it, but it's very worth-while making the point that he or she should do it only for the best possible reasons, namely love of the partner, and better still in wedlock (Holy if possible).

My wife and I are both broad-minded people and modern in the best sense of the word, but at the risk of being called old-fashioned, we both believe in the sanctity of marriage. Why? Because it works as nothing else can. It's all very well being footloose and promiscuous, but it leaves behind a trail of broken lives and shattered nerves. This is what we must make our children realise. They won't be young forever. They can't afford to be foolish forever, either.

As the object of bringing up children is that they should become useful and contented citizens, a parent is duty bound to do everything he or she can to promote true happiness. Today's liberal trend has a basic philosophy of do-what-you-like as long as you don't hurt, or do damage to anyone else in the process. But how can one define the meaning of the word 'hurt' or 'damage' in this context? It's more than physical damage or hurt that is involved. There can be scars that the eye cannot see. No-one can understand another human being fully.

Having seen the experiences of friends who have suffered broken marriages, one wonders if the pleasures balanced the miseries, especially where children are involved.

Maybe the marriage ceremony will eventually disappear, but whether a marriage in name or some other kind of liaison, surely the family unit will remain the keystone on which all society is built. As far as I can see, other civilisations which have abandoned this principle have failed.

Trial marriages get a lot of publicity. On the face of it, the idea sounds fine to youngsters, but what trial marriage could possibly succeed compared to a real marriage, because it is only what it is – a trial marriage. It can't have the same input of effort to make it work, as in a real marriage.

Then, if the first trial marriage fails, do you take a second, then a third, and so on, until an ideal partner is found. It may be too late by then.

Children must be provided with as much technical knowledge for them to form a sensible, well balanced outlook on sex and morals in general, so that they will not be shocked by the sudden impact of some new and strange behaviour.

They must be taught in easy stages, but prevented from ever thinking "Well, no-one ever told me that happened". At the same time it should be impressed upon them that most of the motivations behind these new and alarming developments are commercial, that is, after their money.

If you can prevent the exploitation of human beings, you are on the winning track. Parasitic relationships are the worst kinds of all. Just as the mistletoe grows on an apple tree, or a louse grows on the back of a dog, some relationships use up one's emotional energy, and love and affection, for the selfish gain of another. These are the most degrading of all the many kinds of arrangements.

Responsibility is the most important fact to be impressed upon young minds. While moral restraint should of course be encouraged, one must realise that no one is perfect, and practical knowledge should be part of every young person's equipment. I reject strongly the argument that giving adolescents information on methods of birth control is tantamount to encourage them to sexual promiscuity. Ignorance of the facts is far more likely to lead them off the rails.

So, in short, the advice I give to my children is as follows: "Don't think that sex is just like having a cup of tea. There's more to it than that. Irresponsibility in love affairs leads to emotional upsets, entanglements and at worst unwanted pregnancies".

Now, it's impossible to cut children off from the pressures and the strains that are imposed by the media, yet censorship is not the answer. In any case it's no good blaming the telly, the press, the bookstalls or anything else, or giving pompous lectures on morals, if parents themselves behave with abandon. In other words, example is always the finest teacher, with one exception – that of experience, but this lesson is often a hard one to learn.

It all stands on maturity if people are to accept a reasonable moral standard. Take censorship. If all people were suitably educated and bright, with an understanding of the subject, censorship would be unnecessary and even abhorent. I have mixed feelings on censorship, but I do agree absolutely that knowledge must not be forced upon young people before they ask for it. Do it in easy stages – that's the teaching bit.

Are we right in censoring anything? Well, one doesn't feed a cow with unlimited grass and allow it to eat until it bursts, and it's the same with children. They shouldn't be fed with pornography at an early age. They simply have no interest in it.

Nowadays we are assailed with porn from all sides. It is big business, and to my mind that is the most objectionable thing about it. Now, maybe, some people assume that we derive pleasure from porn, or for that matter prostitutes, or live sex shows and things of that nature. No doubt some people are forced through physical or psychological circumstances to get their kicks at second hand, but porn is no doubt a sign of the 1970s. I suppose in its way it is today's equivalent of Chaucer's writings long ago.

It is possible that some lurid entertainments may cause one out of a hundred people to go berserk, and to be damaged socially, but frankly it leaves me cold.

Whilst on a trip to Germany, a friend and I decided to visit a kind of Ideal Home Exhibition for Sex. Everything was bright, breezy and clinical, and treated with typical Teutonic heavy handedness. One of the features was a live sex show, so out of curiosity we paid our five marks, and joined the queue. Watching a man and woman copulating on a stage, in a room full of middle aged men on hard chairs, was to me, not nearly as exciting as watching orang - outangs feeding at the zoo.

During the performance I wondered if someone had laughed would it have stopped the show? Perhaps the 'entertainers' would have been relieved, as they were billed as performing every $1\frac{1}{2}$ hours during the day. *Tres Formidable!*

The whole concept of this type of thing, although providing a substitute for the real thing with certain types, uses the human being as an apparatus not as a person, and even worse, vested interests are involved, lining their pockets through the weakness of others. Parasites again!

I suppose its all right to watch a blue film every five years or so, in case there are any new developments in the subject matter that one might have missed, but these are generally just variations on the original theme.

If one has no sexual hangups, and has a satisfactory marriage I suppose it's easy to be smug, critical and even supercilious when discussing morals and advising other people. Some people could argue that what's good for one is not good for another, for example they might say that promiscuity is all right for the promiscuous. We all know of examples of people in the artistic world, painters, writers and musicians, who have felt the need to sample all the delights that are available, but I think it's wrong to sample everything, just because its available. If we did, everyone would be rushing to sample not only the soft drugs like cannabis and marijuana, but also that killer drug heroin.

Personally I feel soft drugs are no worse than alcohol. If alcohol had been invented in the middle of the 20th century, it would probably have been banned by now, or at least people would be campaigning for it to be banned. Mahatma Gandhi said "Alcohol ruins one intellectually, economically and physically".

I enjoy a few beers in the pub, but it's a sad fact that many people indulge to a dangerous excess. They must need to get out of their minds by booze or drugs to escape the modern world. I've never seen animals do that, although we once had some pigs who were fed on the contents of some old beer barrels and by gum, when pigs are drunk they're terrible. They go round biting each other's ears, and generally being completely anti-social – just like humans in the same state when you come to think of it. There must be a more economical way of attaining numbness – a wooden mallet would be cheaper for one thing.

I will conclude this little chapter with what I consider to be the ideal state of human relationship. Total involvement with one person, whether he or she be black, white, halt, lame or blind. This brings out the hidden qualities in each partner, which will benefit others who come into contact with the couple.

"This is your life, Ted Moult", the man said. Top: Marie and myself with Eamonn Andrews. Bottom: Mother and Father came too. BBC copyright photographs.

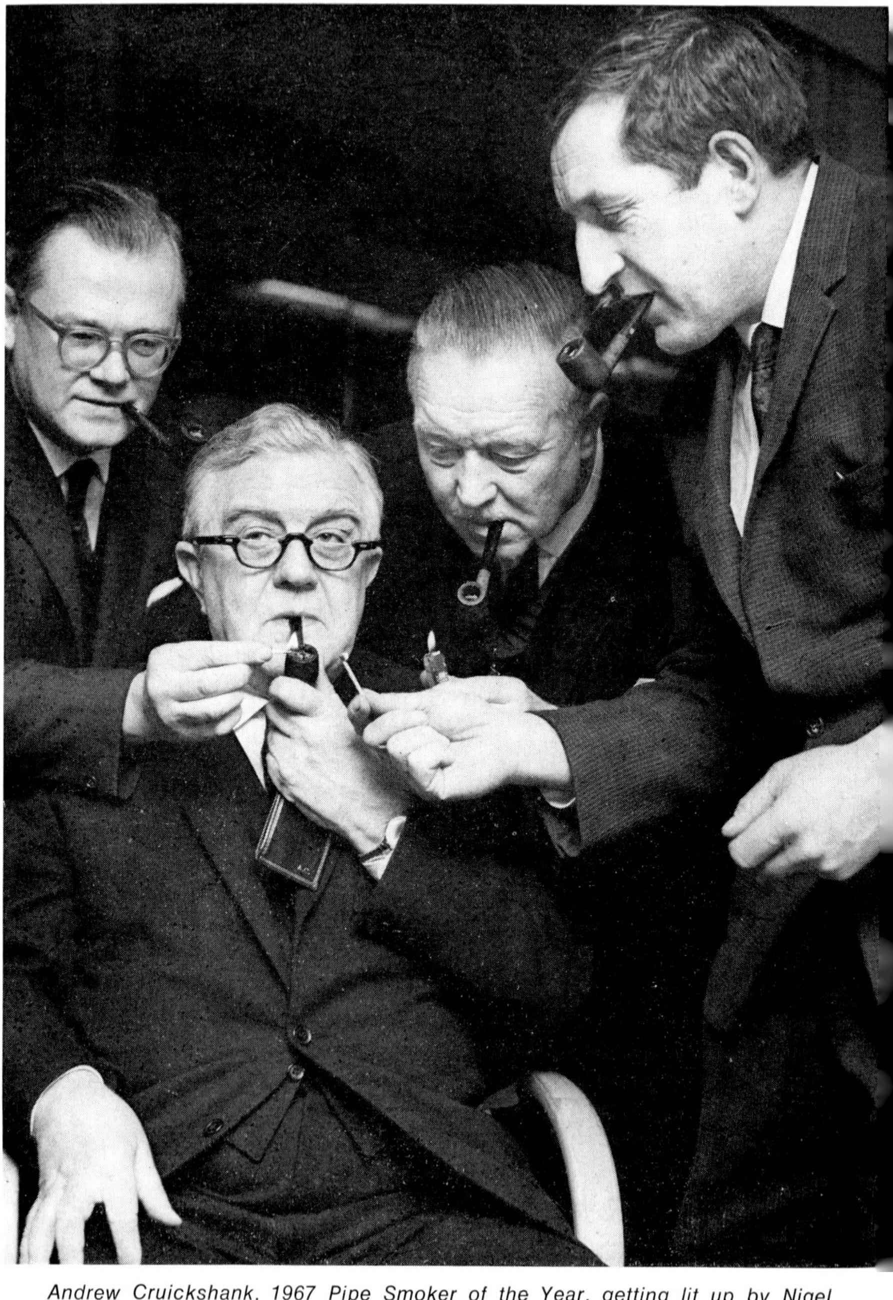

Andrew Cruickshank, 1967 Pipe Smoker of the Year, getting lit up by Nigel Stock, the chairman, and self.

Ashbourne, minutes before the game, supported by committee members Charles Etches, and friend.

Carving an ox for World Refugee Year — pity they couldn't have had a taste!

MOULT IN THE FAR EAST

Above: Self and fellow pipe smoking peasant at Lo Wu, China in the background.

Top right: Ocean Terminal, Kowloon. left to right Producer, Michael Tuke-Hastings, his secretary Brenda Davies, self and Nan Winton.

Bottom right: Trying the bagpipes with the Black Watch in Hong Kong.

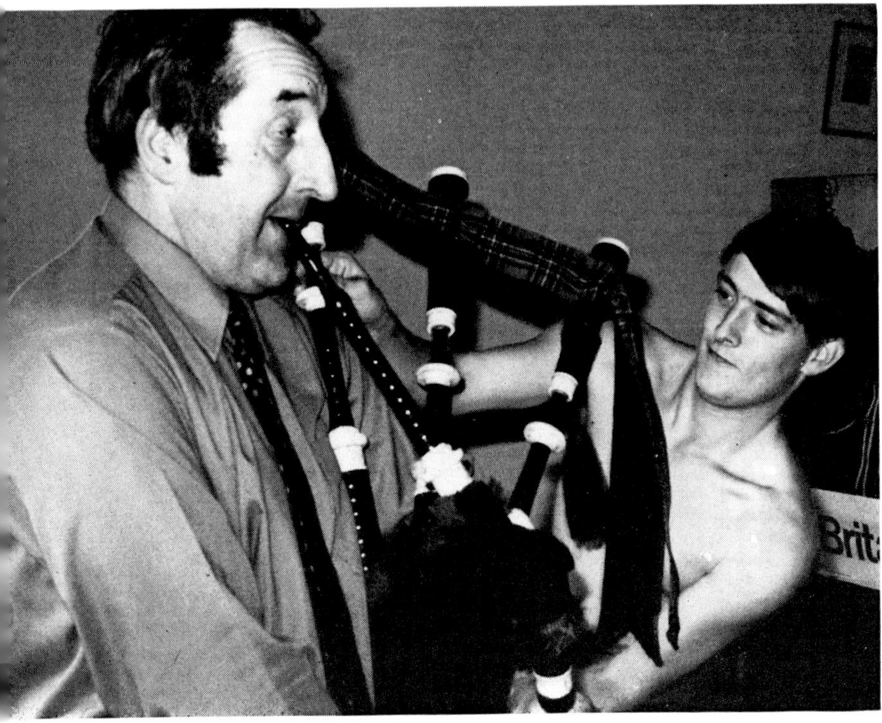

To Denmark. The film unit includes Derek Smith, producer, Peter Caselberg sound, and Ian Stone camera.

Forces Chance in Malta. Back row: Team from R.A.F. Luqa with Brenda Davies. Front row: Michael Tuke-Hastings Alun Williams, Nan Winton, Neil Durden-Smith, self.

With Tommy Cooper at a Manchester Press luncheon.

Playschool takes over the farm. Miranda Connell chats to Anna.

STRAWBERRY TIME

Top : Paying off the pickers.

Bottom : Strawberries can be fun !

13

Farming as a Career

NO MAN can succeed, nor should be encouraged to take up farming as a career, unless he is unstoppable. He must have a positive vocation for the life. Think of the number of hours that make up his working day. A farmer can't clock-on or clock-off; he can't go on strike, or start a go-slow unless he wants to ruin himself. Make no mistake about it: farming is for the rugged individualist of unswerving dedication.

Work consumes an enormous amount of our lives, therefore a man must choose his job carefully. If he works rigidly to a forty-hour week, he still spends the major slice of his waking hours employed in earning a living. Nothing has a bigger bearing on his state of mind, which means health and happiness. Choose wisely: choose carefully! Let everyone realise that farming is only for those who really like it.

In this country the usual career pattern is that a boy takes a job when he leaves school and stays with the labour of his early choosing for the rest of his life. It's not like that everywhere, of course. In Australia, for example, it's not uncommon for a chap of forty or fifty years of age to change from one slot to another. He may go from being a stockbroker to a banana-bender, from an engineer to an interior decorator. The Yanks are like the Australians, and lead more adventurous working lives. There's nothing unusual in a chap taking a petrol station for a few years and then moving on to something else. It appears that in the young countries, young ideas dominate, irrespective of the calendar years of the individual concerned. In a country with a long history like Britain, we're more rigid, less mobile and have a tendency to be more complacent. This is not always a bad thing. There are some people who are round pegs in round holes, which makes them quite contented, even if their job is drudgery. To a placid nature, drudgery becomes a bearable state of existence. Perhaps if they were doing a more exciting job it would create sheer euphoria, or would it bring ulcers? Getting back to the farm, if people enjoy the life, even mucking out or stacking manure can be a pleasant task.

When I started my career, I had the £1,200 in the Derbyshire Building Society, of which I have already spoken. It wasn't a lot of cash, but with it I was able to secure a toe-hold in farming. It's hardly necessary to state the obvious fact that the first thing you need to be a farmer is a farm. I was fortunate enough to be able to rent one. Would this be possible today? Would the equivalent of £1,200, allowing for inflation, be enough to make a start? I think not.

I often wonder how many men took up farming from choice in the old days, and how many because they were expected to ease the strain on dad as he got more physically inept? There was another group whose reasons were suspect, namely those who decided to go farming to get into another environment, escaping the squalor and unemployed misery of the towns. Before the war, in the 1930s, land was cheap. It's hard to believe that there were empty farms going for a song, being let rent-free by big estates just to keep the land from going into decay.

In those days Britain had a large credit balance, and used the granaries of the world to maintain her economic position. It was the policy of the day, the policy of the Government, to have cheap food. I particularly remember reading Chamberlain's speech at Kettering which confirmed this in no uncertain terms. It certainly looked as though the luckless farmers hadn't much future at all. Wheat was as low as eighteen shillings a quarter; farm labourers received about thirty-two shillings a week if they were lucky; a good dairy cow, an Irish in calf, or a short-horn heifer, would fetch from eighteen pounds to twenty pounds.

I vividly remember that the Daily Express, to their credit, hammered home a pre-war campaign to raise the level of agricultural wages, morale and efficiency. They made their point not only in the news columns, but also in a subtle and humorous way by regularly featuring a cartoon character portraying a rustic chewing on a straw. His name – 'Idle Acres'.

Things got so bad that the better farmers were the ones who suffered the most. Those who could cut a few corners, even if they didn't cut their hedges, weathered the period best. The overall effect still exists in older farmers today, in that it has given them an acute dislike of parting with cash. It has made many of them mean and parsimonious. Faces grew narrow in those distant days. Many were so embittered that it has actually stopped them expanding from that day to this for fear of failure. Farmers who stuck to the essentials survived, but large numbers went out of business, especially in the eastern part of England. Fortune favoured the livestock farmer, however, particularly the chap who milked cows. This was largely due to the creation of the Milk Marketing Board in 1933. All farmers dealt with this Authority which regulated prices between one region and another, so that one farmer and another levelled out. It was a fair deal, fair shares. The certainty of receiving a monthly and regular pay cheque gave a degree of security hitherto unknown to such farmers. Today it is taken too much for granted. Are they better men for it I wonder?

The milk cheque became the farmer's equivalent to the wage packet. If a farmer owes you money, it's still a good idea to ask him for it when you know he's expecting to receive his milk cheque: it's around the 20th of the month!

Those who are thinking of going into farming cannot do better than consider in depth the real problems that existed a few short years ago. As I've said, things are easier now, but it's still a dedication. In the pre-war years, family farms were the order of the day. These survived by and large because they supported the whole family, under the system that every member worked gratis for father, with a vague promise as far as the sons went, that they would get a tenancy farm in due course. This I feel, had a bad effect on father/son relationship. The son worked to be around forty, and in many cases was still asking dad for sixpence for a haircut. Often, a tyrannical father never let go of the reins, and I know one case where such a man married twice, having worked the first family and wife to death, literally in the latter case. The result was that the embittered first family saw the father and the second wife enjoying the fruits of their years of labour.

In the old days the bright lads often left the land. Things got to such a pitch that if a farming family suddenly realised that they had a bright offspring, they would have a financial whip-round to educate him or her, so that he or she escaped and became a school-teacher, a doctor, or some other highly regarded member of society. This is why education to this day is stronger in the poorer districts. It means more, and is valued more. The end result of all this was that those who weren't very bright were left on the farm. How things have changed!

Many of the improvements in farming, and the best farmers, were a direct result of a change in thinking habits. Some farmers faced the job with a new approach and questioned the reasons for doing things in the old order. These men wouldn't accept the traditional style without close scrutiny, and subsequently they were proved right.

Before the war seventy per cent of the farm land in the United Kingdom was occupied by tenant farmers, most of whom rented from big estates as I've already explained. Some estates looked after their tenants in a quaint patronising way. Others just didn't care. The worst examples were those farms that existed almost exclusively for the pleasure that could be derived from them by the rich land-owner, who was usually interested in little more than the hunting and shooting aspects. To secure a farm for his son, the poor tenant had to keep on the right side of the landlord *and* the gamekeeper, and too often, by the time the son finally got his farm, he was getting on in years. Moreover, it was a case of "Well, son, take twelve cows out of the bottom shed, and buy a mowing machine and some hay-making equipment from a sale, and you're in business". It was a challenge of spartan proportions. Many cynics would say it produced better men than the softer system of today, but how does one judge?

Some interesting statistics emerge from the pre-war days. Only a third of the cheap food needed to feed a population of forty-eight millions, came from British farms. Consequently the small farmers had little confidence to match. Theirs was a hand to mouth existence, with minimum incentive to get bigger and better. The markets were flooded with cheap food from overseas, and home crops fetched a pittance.

Today, we produce two-thirds of what we eat, discounting things like pineapples and tropical fruits, of course. This change of fortune occurred during the war, and provides one example of good coming from evil. War-time conditions revitalised farming. Milk became a number one food and has maintained its position ever since.

When hostilities ceased, Britain was left with empty money-bags, and consequently no longer had recourse to the resources of the world. Britain always needs her farming industry when she's short of foreign currency, or for strategic reasons. Today, with a weak currency, we are experiencing a similar dilemma. Added to that, present world prices are high. The farmer is again in demand.

On top of this, available farm land is 'shrinking', partly due to the population explosion. It's been estimated that today we only have, as individuals, three-fifths of an acre for living, working, playing, and producing food. Compare this with France, where there are two and three-quarters acres for each man, woman and child. Here lies the hub of the problem. More mouths to feed, and less growing area for essential food stuffs.

Having said all this, let's get back to our central subject. People regularly ask me what opportunities there are for their son in farming. They seem to imagine that if their young hopeful gets some kind of qualification at an Agricultural College or even a degree, he has arrived, and the son invariably has similar ideas. He sees himself riding about in a Landrover with a hunting and shooting stick close to hand. This dream picture must be dispelled instantly. There are plenty of would-be farm managers. What is needed, are fewer managers and more useful operators!

In my opinion, the modern farm worker is top of the industrial pops. That's one view that hasn't changed with my widening horizon. The evidence for it is undisputable. The farm worker can find himself alone in charge of a hundred cows. How many people realise that they are worth £25,000? My last combine-harvester cost over £6,000. The point I'm making is that today's young men on the land carry a heavy responsibility in all senses of the word, including financial. They must be resourceful, capable of making decisions, and decide when to stop and start a particular job. This is governed by a number of factors, not least of which is the weather. Contrast the farmworker of the 70s with his predecessors. Today a couple of men with machines are doing what teams of peasants did in the olden days. Men are restricted today by the limit of their brain power, not by the size of their biceps as of yore.

If the over-riding consideration of the young fellow coming into farming is that he wants to be his own boss, he is likely to be disappointed. It can't be repeated often enough that farming is a dedication, not an escape route, or an easy method of working for yourself. The entrepreneur would be better to take a pub, or run an antique shop. The essential qualities for working on the land can be summed up as follows:

1. Love of farming.
2. Business ability.
3. Labour management.
4. Getting up early in the morning!
5. Big thinking.

The latter point needs expounding. With each passing year, running a smallholding is becoming less and less attractive. Take, for example, those who keep poultry or pigs for commercial reasons. Their livelihood is being eroded as a result of these products being produced by large organisations on a vast scale. The 'big boys' have the advantage of bulk-buying, and lower unit cost comes into full play. This results in a product being cheap enough to be accepted in the competitive world of marketing. What then is the answer for those entering the 'arena' armed only with limited resources?

The increasing trend is for the small men to get together. Instead of four farmers each keeping fifty cows, it is possible for them to set up a combine, comprising a two or three hundred cow unit. They can then appoint a manager, it may be one of the team, with each of the others playing the part for which he is best suited. I also see a future for share cropping. Let me make this clear. A man of some means may not wish to part with his land, but may be willing to bring in a young man to work it and share the profits. This is against the English tradition of an annual rent being fixed, but times change. When I was in America I stayed on a farm where the landlord provided half the fertiliser, half the seed, and took half the profit. It seemed a good arrangement for the landlord, but hard on the tenant by British standards.

I have touched on the subject of marketing, and the importance of understanding it. A successful farmer nowadays has to be a clever chap in the wider field of life. However, none of us is so clever that he can completely disregard Mother Nature. The effects of weather are still unpredictable, though less significant than they used to be, thanks to mechanical aids like corn dryers, and modern equipment in general. More work can be done quickly now and at the most appropriate time. Timeliness is the key word for success in farming. Knowing what, where, and when is vital, but at the end of the line the goods still have to be sold. Very much so! In fact, it's harder to market and sell than it is to produce a basic yield. Marketing is a science in its own right. Buying and selling is not my favourite job. Dealing in cereals, for instance, with the recent wild fluctuations in price, due to a market sensitive in the extreme, is like trading shares on the stock market. Greed and fear (both unattractive qualities) govern both activities.

If you *must* become a farmer because of an inner urge, good luck to you. For some there is no other way of life, but this is no consolation for the thousands of young people who want to get into farming from the outside.

How do they find their niche in this big industry, and at what level? They know what's desirable, but what's possible? Well, the chance of farming on their own account, is, I'm sorry to say, very, very small. The best thing they could do, if it's not winning the pools, is to marry the richest girl, or the girl with the richest father, in the Young Farmers' Club.

The main reason is the cost of the basic raw material – the land, which has shot up in price like a rocket from Cape Kennedy in the last 5 years, and accelerated violently in the year 1972 to 1973. Land is so expensive now that if one could afford to buy, one wouldn't need to work for a living, and if one bought it on borrowed money, one wouldn't make a living anyway with the high interest charges on the capital. Land with its special qualities has become a desirable object for investment, and while the freehold system lasts I think it will continue to attract more capital because of its relative security than for its other assets. In fact, one can even buy agricultural land bonds as an investment.

As I have said in another part of the book, land is disappearing from the scene at the rate of about 60,000 acres a year, and anything that is shrinking and rare, like the best reserve grape, fetches a high price. Another reason, of course, is the legislation which enables a farmer, or for that fact, any landowner who has held land for a period of twelve months, and sold at a high profit for development, to invest this money back in land within a given period, thus postponing the eventual payment of Capital Gains Tax. So if inflation continues, I suppose land will go with it.

A curious side effect of the increasing value of land, is when a farmer has willed his possessions to a son, at his demise. The death duties could be crippling.

Any farmer thinking of dying, (and that's not so rare as you would think), should have given the farm to his heir years ago. However, old drivers don't chuck away the reins so easily, and in any case he may fear something happening to the son. So you get the situation of a farmer with the same earning cacapity as before, yet with an added estate value, owing to the higher market price of his land. When the old man kicks the bucket, a lot of money has to be paid into the tax man's kitty, even allowing for the 45% estate duty rebate and although his son takes over the farm. The only person who higher land prices really benefits, is the chap intending to retire.

Whatever the reasons, though, my main object is to point out the improbability of buying a farm from scratch, and to make a living out of it. Although, let it be said if the embryo farmer already owns some land, say 400 acres, it would quite sensible of him to purchase another 100 acres, thus averaging out the cost per acre.

When discussing return on the capital in farming, I consider it important to distinguish between the tenants' capital – the tractors, the cows, the combine harvester, the growing crops, stocks of hay, sileage, straw, sugar beet and so on – as against the landlord's capital, even if the farmer is the owner/occupier, and is, so to speak, his own tenant.

What about renting a farm? Again the chances are very slim, because of the high value of land. For every farm that becomes available, there is a long queue of potential occupiers. The practice of letting them to the highest tender, also results in the rents being reasonably high, although a sensible landlord will charge a rent which will still allow a tenant farmer sufficient financial return to keep the property in good order. Rents usually follow the value of the land on the open market.

In any case it will surprise me if landlords fail to see the advantages of farming their own land – 'In hand,' as they say. Even if the net profit is no higher than the letting value, the landlord would avoid the 15% surcharge on unearned income, which rent is, of course.

The returns may seem abysmally low when considered against the value of the land being farmed, but quite healthy in relationship to tenants' capital. As land is a safe appreciating asset in any case, one shouldn't expect big returns in this respect.

At this stage let me point out that my intentions are not to frighten people away from a farming career, although observations so far may indicate so. Contemplating a career is as important to a man as selecting a wife, if he is to be successful, and one must weigh up the pros and cons objectively.

Unless he is well breeched financially, a man with a burning desire to work on the land must resign himself to the fact that his labours must be carried out on someone else's farm. He must be an employee, and here again we hit upon one of the 'cons'. Unlike many other industries, the management structure is fairly rigid. As farms are relatively small units, the promotional ladder is not the way to the top, as in big public or private companies.

On the other hand, in farming one has the opportunity to show initiative and responsibility, and to work as a member of a closely involved team. To some people, these attributes are worth all your promotion prospects, and there's no need to watch your back on a farm. Not from workmates anyway, but the unaware can receive a nasty shock from a playful animal!

How, then, does one embark on a farming career? Most counties have a College of Agriculture, formerly called Farm Institutes, and although some of these require a student to have one year's practical experience on a farm as pre-college experience, some prefer to take a boy straight from school. The latter have the best system, in my opinion. The student is taught all aspects of agriculture, so that his particular aptitude or preference can be identified. A young person who has previously worked on a farm may have spent his time exclusively with one type of stock, or restricted by the types of crops grown. Many farmers are so specialised that initially they may have a narrowing outlook on a learner.

There is an apprenticeship scheme for farming aspirants, but I must say that if it is not a flop, then it appears so, since only about 15% of eligible lads enter. It also appears beneath the attention of farmers in general.

Careers Masters and Youth Employment Officers appear to lack an appreciation of the qualities required in a farm worker. Dull, slow chaps are better suited to the mass production lines than on our farms and in our fields. There is a definite need to improve the image of the modern farm worker in the eyes of the general public, especially as some of the prospective new recruits will come from the cities.

I have noticed over recent years, that the present farm worker and the farm worker of the future is by necessity a totally different type to that generally visualised.

The drift from the land which has occurred for the last 20 years, and has taken a quarter of the farm labour, has been by tacit agreement, more or less, on both sides. Farmers replaced hands by horse power, and this has proved the right policy. Obviously, a man is as good as the amount of horse power at his elbow, yet the cost of machinery has risen to such a high level, that one wonders if the drift from the land can ever continue as in the past. Fortunately, there are signs of it stopping, but in my opinion, it has been reduced as low as it can be. Tractors can plough fields and combines can cut corn, but they can't recover straying cattle. Hands are essential on a farm, and I mean hands in the literal sense. Agriculture has been a reservoir of labour, from which any new industry could poach chaps for the factory, but I feel that farmers have accepted the reduction of their manpower, streamlined systems, and taken on, for instance, single enterprise types of cropping, which make the use of labour very efficient. At the same time, this has enabled them to sell off a few cottages.

Farm labour is *pari passu* with industry at long last. It certainly deserves to be, because the modern chap working on a farm has to be independent, bright, mechancially minded, and observant. The days of gangs of hoers and mowers have long past, and because he will be working on his own, he must be self-sufficient. By self-sufficient I don't mean the ability to take enough sandwiches and coffee to last a day on the tractor, but the ability to make instant decisions, if conditions change.

The pattern in the old days was that many youths joined farming at an early age, and after a few years abandoned driving a tractor in the fields for driving a lorry on the roads. Today's trend is to keep these youths on the farm, and to give them an equal life to those driving lorries. No doubt about it, farming is a skilled job and it takes a long time to learn. One has to get the feel of things; to know when the hay is ready, when the ground will take a potato planter, and when the ground it too wet for ploughing. Ploughing under the wrong conditions would ruin the soil structure down below and probably cause damage lasting several years. A stockman has to spot the pig before it gets really sick. If pigs had measles it would be quite easy, but one must have the ability to identify an animal who *might* be sick, and to isolate it immediately, because disease is a crucial factor in livestock husbandry.

Possibly the best way for an aspirant to start his farming career is to join a Farmers' Relief Service. This is a splendid scheme in which farmers who want a week's holiday, or break a leg, can arrange for a stockman, herdsman, tractor driver or even a foreman, to carry them over the period when they are away or otherwise indisposed. In a relief service one obtains a great deal of experience in the shortest possible time. However, this is only one of the alternatives for a go-ahead young chap.

One of the more recent developments in farming is the situation where large companies take over the running of a farm, probably leaving the ownership to the original farmer who wishes not to part with the land, but still to have a share in the execution of the agriculture. These firms require very bright operators, and the prospects are good.

Whether a college or a university education will help gain a position is a moot point. In my experience, undergraduates are in two categories – those who know what they want to do and take a professional degree at medical school, or in engineering, etcetera, and those who have no fixed ideas, and take a degree in some subject they find attractive, like ancient history, philosophy, politics or even law. Thus they put a barrier of time between their schooldays and the awful day when they must decide how to earn their living. No-one should be pushed into some horrible career that he doesn't want, and many people are uncertain of their working future, until they have reached some kind of maturity.

University life has its benefits, but having visited 16 of them in one year during recordings of the *Treble Chance* programme, I found many of them full of students who apparently wanted to be just that – students in perpetuity. No doubt some of them would eventually become housed in academic life as lecturers.

The achievement of an honours degree lower than a First does not have the same distinction as ten years ago. Having obtained the written evidence that, on a certain date, the examiners were satisfied, the graduate must then learn a craft or skill. An unfortunate fact is that many persons leaving universities have no idea of how to deal with other humans, let alone command any of them.

I feel sorry for many of today's kids. They start on the rat race almost as soon as they become literate, often with parents breathing down their neck, urging them to strive to greater heights than father reached. To stand the course with our stupid insistence on O and A levels, and then to carry on with three more years' study after school is too much.

Employers are often to blame for the perpetuation of the examination system. A bit of paper is the all important aim, and this can be achieved without any *real* knowledge of the subject.

My father-in law coached a young man in Latin, to pass his O levels in three months. The student knew no Latin at the start, and little more at the end, but he passed the exams. One boy passed O levels in Religious Knowledge and Carpentry. Presumably the Careers Master suggested that he would be a smash hit as an undertaker!

Potential undergraduates might benefit from their time at college if they stepped off the endless belt of the education system for a year or so. At least they would see how the rest of the population live, and act accordingly to this experience. I often wonder where agricultural graduates go to these days. Previously they joined the Advisory Service, and ancillary research department of various commercial firms, but I suspect that many students who go on various degree courses are really farmers' sons who plan to take over from Dad in due course.

If your hopes lie in farm management, it's little use asking if it's a worthwhile career. There are so many categories that one might as well ask "Is it a good job being a football manager?". As farming units get bigger, there will, of course be some super jobs available, but, by the same reasoning, they will be few and far between. At the other end of the scale, the man in charge, perhaps working for a pop star or a business man with an interest in the countryside, could be no more that an ordinary farm worker. To him falls the responsibility, with perhaps insufficient reward.

There is a need for well-educated young man on the farm, but he must be on an equal footing with men in other branches of industry, sharing in proper pension schemes, with perhaps a carrot of shared profits. This, I think, is most desirable.

There's a lot of guff talked in the newspapers these days about plans for, say for instance, a National Workers' Fund as a step forward to workers' control. It's hard to see how any industry that includes boring jobs like fitting bumpers to cars on an endless belt, can ever justify a scheme for workers' control, from either a financial or an executive point of view. As anybody knows, counting heads has a limited application as far as any decision making is concerned, but there is a real value in any scheme for industrial democracy whereby the workers can take part in decisions which affect their livelihood. Not just having a share in the profits either.

Here I think farmworkers have a feather in their cap because any sensible farmer, although his is the ultimate decision, will ask Jack's advice and Bill's advice, and doing so gives the chaps he employs a chance to rise to responsibility.

I have found it very encouraging that most people do rise to responsibility. The tragedy is that most workers in this country never get the chance, and they don't get a rise either, not of that kind anyway.

People employed in agriculture have a highly developed sense of purpose. They usually know what they are doing, why they are doing it, and can see the results of their labours. After all, if you give a bucket of barley meal to a pig, at least you can see him scoffing it, which is more than one can say for some of the boring, repetitive jobs in industry.

14

Livestock from the Factory

IF THERE is one controversial practice, at which opponents speak out with apparent sincerity, yet blatantly show double standards, that practice must be the intensive rearing of livestock. In other words, factory farming.

We've all heard the arguments time and time again, that animals or birds should be allowed to live natural lives, that it's cruel to keep them penned up, purely for the benefit of the consumer. Then on the other side, we have the 'anti', whose thoughts venture little further than his table. He will assert that meat, chicken, eggs are not what they used to be. "It's all this artificial rubbish they're fed on. Why don't they let 'em out to root in the field", is usually his soapbox.

Well, it is possible that under ideal natural conditions, food would taste better, but what are natural conditions? It would be natural for a hen to lay a dozen or thirteen eggs like a pheasant, which actually would manage a few more, go broody, scratch around, then retire until the next year. It would be natural for a pig to have huge tusks, huge shoulders, no hams and rush around jungles minding its own business and independent of man. It would be natural for a cow to produce milk and one calf. Man is the culprit. He has decided that animals are one of the raw materials for his exploitation, and has bred them over the centuries, bent on his own survival, and to add variety to his table. The result is that we have hens that lay twenty times more eggs than they would in the truly natural state, cows that give ten times more milk, and pigs that bear hardly any anatomical relationship to the old wild boar from which they descended, but which are able to fill ham sandwiches.

Unless a person is an out and out vegetarian, he's got to accept that unnatural conditions must prevail to produce the muscle, milk, and eggs which most people seem to require. In the dim distant days of history it didn't matter how many eggs and how much bacon the animal produced, but with the population explosion on the increase, today's humans gobble up everything. Is that cruelty to animals?

It's the same with arguments about the battery system of poultry keeping. Today nearly all chickens are bred by this method and kept indoors for the duration of their life. Who dictates this policy? The 'wicked' breeder? I'm afraid you can't get out of it as easily as that. It's Mr. and Mrs. Average to blame. They want lower egg prices and demand them. They like to go out in the evening and have chicken in the basket at the local pub or restaurant. It delights them when the price is reasonable. Yes – both eggs and chickens provide a nutritious but economic meal. In my opinion, eggs are still the cheapest and most sustaining single food item we can buy. But what about the mental blockage of those members of the public who are pleased with the edible advantages, but raise hands in moral indignation at the living plight of their gastronomic victim? I doubt if they'd be quite so voluble if it hurt their pockets.

It would be more sensible of them to realise that eggs and poultry wouldn't be such a good buy if the birds were ranging around in the open air, subject to all the elements and uncontrolled diets that this life-style entails. In addition, the humble hen would be in even more danger from foxes and other predators. If you doubt my words, compare the price per pound of a free range old boiler against that of a tender battery fowl.

Personally, I have no particular liking for battery farming, but I do realise that without it; that is, without scientific breeding, exact nutritional requirements, and controlled environment; eggs, poultry and certain meats would be at least double the price. Being also human, I appreciate the objections that most men and women have regarding the ethics of the system.

When the British Egg Marketing Board was in existence, it tried valiantly for years to overcome another common prejudice. Namely that eggs with a stamp of them were less desirable than those fresh from the farm. This is a fallacy, but one they could never eradicate. They even went to the extent of employing 'yours truly' to help change the public attitude. I appeared at functions during what was known as Egg Week, and in demonstrations cum exhibitions up and down the country. It was no good. The public just wouldn't accept that a rubber-stamped lion had a place sitting on an egg. The lion and the unicorn might sit together in heraldry, but the lion and the hen – never! To many folk it meant that the egg wasn't 'farm fresh'. They had a hang-up from their war-time experiences when stamped eggs were associated with rationing.

Although it was scientifically proven that eggs bearing the lion mark were as good as could be bought hot from the stack-yard, and that there were no better eggs at all, women remain unconvinced. Who could hope to survive against set feminine opposition? Certainly not the Egg Board. It has gone.

What's at issue here is not eggs, but human prejudice. We're a bullet-headed lot sometimes. I don't believe anybody could tell the difference between a lion-stamped egg and one they would term new laid. I find it the same with wine. Perhaps my palate is atrophied or something, but I recognise other people's ability in this respect. There's a lot of nonsense talked about the year of a wine, but I can't tell the difference between a '55 and a '65 Burgundy. At least not until I look at the label and I suspect that I'm not on my own in this. Still, public prejudice can be profitable to unscrupulous people. Some years ago, I recall a chap who made lots of money out of an original fraud. He found a way of getting rid of the stamp mark on eggs. Without the lion, the housewife thought the product so much better, and the taste quite different too! The moral of this story is to make sure you're not conning yourself.

However, I transgress. The subject in hand was factory farming. To understand why such a system has been developed, one should first admit that farms are not intended as havens of rest for animals or birds, and that crops are not grown for their aesthetic value. Farming is a business, and therefore just as subject to increased productivity methods as other forms of industry.

These recent developments, which have caused such a furore, ensure that livestock can perform to the best of their ability, the job they are bred for in the first place. That job is producing food for the human race, and their optimum efficiency is obtained by housing them for life, in controlled conditions.

Is this any more cruel than the Medieval system when farm animals kept themselves alive by scavenging? No crops were grown specially for their food, and very little fodder was conserved for the winter. Cattle and pigs were simply herded into the woods and common land to fend for themselves, while the poultry scratched a living round the farmyard. Came winter, great numbers were slaughtered and salted down, and but a few kept alive for next year's planning.

Consequently any type of breed improvement was restricted, until larger crop yields eventually allowed animals to live longer and to reach maturity. At the same time, better nutrition produced a larger animal.

Not the simple chap as generally pictured, the farmer realised that he could improve the quality of his stock by suitable breeding, and if he adopted better methods of feeding, housing and disease control, production of meat and eggs would increase accordingly, and fill his cash box that much quicker.

It may surprise the reader to learn that factory farming is not entirely a new development. Take the old town dairies, for instance, which were considered admirable institutions not so many years ago. These were probably the forerunners of today's system. Fresh milk was produced on the spot for a growing urban population, but to achieve this, it was necessary to keep the cows indoors, and to bring in their food from outside. Town dairies flourished for many years, without opposition, until the development of the railways enabled fresh milk to be brought from the country more cheaply.

This idea of bringing food to animals instead of allowing them to forage or graze is one of the main advantages of intensive husbandry, and is becoming more widely adopted as farming units become larger.

Nowadays, on large scale units, the very best use is made of the three essentials of all farming enterprises – land, labour and capital. Often these units are absolutely self contained, one organisation looking after breeding, rearing, feed purchasing and compounding, and marketing the end product. Even the waste products are not entirely 'waste'. This is turned into dried manure to enrich soil on which crops are grown and could even be re-cycled as it still has some energy value.

Sophisticated equipment is installed for automatic cleaning out, egg collection, insulation, ventilation, light and humidity control, but all these would be little use without adequate health precautions. In a factory farm, one infected animal or bird could create havoc within a very short time, so control of disease is imperative. Modern vaccines, drugs and sprays have made this most effective in this day and age, but one curious side effect that came to my notice, is that certain ailments which were previously obscure, are now becoming more widespread. Perhaps Mother Nature is showing that she is still more powerful than Man, whatever his technology.

All farmers have long been of the opinion that disease control precautions should be carried out on a national rather than an individual basis. The fairly recent outbreaks of foot and mouth disease, swine vasicular fever and fowl pest have added weight to our arguments. With more animals and poultry kept in fewer hands, there is a better chance that this will eventually happen.

Factory farms concentrate poultry, including ducks and turkeys, and animals such as pigs, rabbits, sheep and cattle into a comparatively small area. Consequently, all aspects of their development can be better controlled.

Restriction of movement is one of the factors essential to the efficient production of food, and this is the practice which most upsets the feelings of opponents of the system.

The production of veal, a very small percentage of the total British meat production, offends most humanitarians.

It's hard to understand why calves should be slaughtered at such a young age when the newly born calf is so expensive. It would provide far more meat on the same skeleton if it survived to become a two-year-old.

However, to satisfy the demand as it is, calves are penned for their lives, but normally in hygienic conditions, in a house maintained at an even temperature. They are fed on a liquid feed of high nutritional value until they are 13 to 14 weeks old, and at an ideal live weight of 350 lbs. They are then a marketable commodity.

Opponents of intensive methods on humane grounds are strongly critical of any system which fails to provide room for animals to get up, lie down, stretch their legs, (or wings in the case of poultry), turn around, or to groom themselves. However, a committee headed by Professor Brambell, published a report in 1965, which recommended better conditions, but did not consider them essential. The committee considered that the majority of farmers were concerned in the health and welfare of their stock, and didn't condemn their attitude in general.

Later, in 1970, a survey by the Ministry of Agriculture, Fisheries and Food veterinary service found no cruelty attributable to pen width, in veal production. I wonder if it's a case of what the calf never had he never misses!

We never hear the same outcry against fish factories. Centuries ago, the monks kept a well stocked carp pond for their meatless obligations. Nowadays, with greater understanding, fish breeding, incubation and rearing is carried on intensively under factory conditions in special tanks. It would be amusing to see the public's reaction to a 'Cruelty to Fishes' campaign.

Whatever our views on the ethics of factory farming, productivity is the criterion. Progress has its price and everyone must pay. In this case the price is lower than it would be otherwise.

Without factory methods, I am sure eggs would be in desperate short supply, and chicken would be a luxury for special occasions. With the breeding cycle of poultry being relatively short, genetic progress has been rapid, and the modern hybrid hen has far more uniformity than its ancestors. With precise management of feeding and environment, a production of 320 eggs per year from each hen is predictable. A far cry from the free rangers whose production was, to say the least, a little erratic.

When man was more primitive, hunting was an essential activity. It provided him with the means to secure food and clothing. Prehistoric man covered himself with skins after eating the meat of his prey. In spite of all this, the animals suffered less in decline of numbers than they do today, and were free to roam over the entire globe. In this century they are sadly compressed to ever-shrinking areas, which make the survival of certain species difficult, if not impossible.

Humans are today so thick on the ground, and take so much space with their cities and sprawling villages, that they're driving the animals into a corner. The thought also occurs to me that having thus intensified the animals into limited spaces, we shall soon have intensified humans to match. What's the difference between high-rise flats and battery cages? The most depressing sight I saw in America was during a visit to a small TV studio on the West side. Through a window I looked out onto the housing conditions below of the Puerto Ricans and the Negroes. They appeared to be living in what were humanised filing cabinets, with spidery fire escapes threading their way twenty or thirty storeys down to street level. Are we becoming battery animals? Is such incarceration in huge cities going to be our ultimate fate? Gandhi had a simple approach, Diogenes lived in a tub. I wonder whether we will eventually be able to choose for ourselves. Even in the USA the ideal way of life, as eulogized by the Founder Fathers of that country, was a hundred and sixty acres or a quarter of a square mile living space for each individual, coupled with yeoman independance.

Can we really condemn intensive rearing of livestock developed as a result of public demand for its product, when human beings may eventually suffer a similar fate?

15

Life and Leisure, Country Style

RURAL life is not only work and sleep, but for some it used to be almost that. Looking back I'm amazed at the change in life style during the last 30 years.

In those days, no one seemed to have any time off. Certainly not in farming. When I started my labouring career the only statutory holidays allowed to farm workers by the magnanamity of the Agricultural Wages Board were three days a year, plus the usual Bank Holidays. Since animals deign not to recognise such non-essentials, even our free days were occupied, unless someone volunteered to milk the cows, feed the pigs, and carry out the essential daily tasks that befall every farmer.

For the entrepreneur on a small scale, the situation was even tighter, and many a small farmer took it as a matter of course to spend year after year on his holding without relief. To provide a break from his daily chores he would look forward to the local agricultural show, when he could snatch a few hours relaxation, or, if very lucky, a whole day out.

You may consider this a busman's holiday, but in this respect farmers are very much like publicans. All the publicans I know spend their evening off in another pub, and I bet the 'mine host' of the reader's local does the same.

One small farmer in my own county was reputed never to have left Derbyshire until he accompanied the local Farmers' Union on a day trip to London. He wandered through the metropolis completely unaware that some wag had pinned a notice on his back, but this notice read not "kick me" nor anything as puerile, but "I am Billy Goodall from Ashbourne". As he heard the passers-by remark "Look, there's Billy Goodall", he wondered how he, who had spent his life with mud on his boots, could have become so famous.

In Derbyshire we seem to breed the 'characters', but I feel the same could be said for any district where the farms are small, and the land is difficult, or, in farming parlance, 'hilly'. The influence of change is limited, and thus the culture more enduring.

"The mountain sheep are sweeter,
But the valley sheep are fatter.
We therefore deemed it meeter
To carry off the latter".

That snatch of poetry refers to Wales, but is equally applicable to the upland district of Derbyshire, where they turn the cows out to graze at least a month later than the more fortunate bovines in more favoured climes.

The difficult land of Derbyshire is either on millstone grit, home of the Derbyshire Gritstone breed of sheep, or on limestone, a region which merges into Cheshire and North Staffordshire. I once heard a parson giving a sermon for which the text was "When a sheep bleats it misses a mouthful". I never discovered its origin but I thought how appropriately it could be applied to many situations, or even personalities for that matter.

Ashbourne, one of Derbyshire's most attractive towns is in this area. Self styled 'Gateway to the Peak', its main claim to fame is not that it's the birthplace of Billy Goodall, but for a lunatic game of free-for-all football, which erupts every Shrove Tuesday and Ash Wednesday.

The game is a very old custom, with its origins, rumour has it, in the delightful practice of kicking a Danish warrior's head up and down the main street in the dim and distant past. Whatever its belligerent antecedents, they are undimmed by time. If one can imagine a combination of an All Black loose maul, hunt the thimble and a riot in Calcutta, that's the Ashbourne football match.

The teams, of unlimited numbers, are divided into those born north of the Henmore Brook, which flows through the town, and those born on the southern side. The site of the local Maternity Hopital is disregarded. I suppose disputes could occur as to whether responsibility for its exeants lay north or south of the brook.

A fierce partisanship exists, handed down from father to son, and no quarter is given or expected between the two goals, which are mills, three miles apart.

The proceedings commence with a traditional luncheon at the Green Man, and after the last draught has been quaffed, an invited celebrity is carried, with a gaily painted ball to Shaws Croft. Then after an exhortation to the multitude to play the game, keep it out of the churchyard, and not devastate the town completely, the National Anthem is sung, the ball 'Turned Up', and battle commences.

For the first few minutes there's a fair amount of disorganised kicking of the ball (and each other) while the two sides warm up, but before long the ball lands in the brook to start an unrelenting struggle between the dedicated. The locals call it 'The Hug'. Sometimes the ball emerges quickly, but more often is buried for ages below a melee of struggling townsmen. I once spent four hours in those icy February waters, and one could see the steam rising like a mist from the knot of scrummagers.

Whoever manages to get the ball to his opponents' mill is acknowledged a local hero, and has the great honour of being allowed to keep the ball.

If there is no score the ball is given to the personality who started the game. That's how I got mine, which is, I suppose a bit of a cheat. However, hanging from the ceiling in our kitchen, it provides a good talking point if conversation flags.

With communication somewhat limited years ago, the social life of a community was more self-sufficient, and revolved round the functions held in the Village Hall or local school, with the desks pushed to one side. As one sniffed the atmosphere peculiar to all classrooms, one hoped that one hadn't supped too much beer to remember the steps of the Valeta, Military Two Step and the other old favourites, now mercifully preserved on *Come Dancing*. Accomplishments in these intricate steps was a 'must', both for the married man if he was to please his lady, or for the young blood impressing the local female talent. Nowadays, it appears to me that the only requirement to be a success on the dance floor is a distinctly individual approach.

With the endless games of cards, mainly halfpenny nap, darts, dominoes and the like, we filled the need for diversion in the winter months. Summer, of course, was the time for work, but there was always time to knock off the odd rabbit in the pre-myxamatosis days.

Most of our travelling was by bicycle. At one time I lived on a farm which had a one and a half mile drive across the fields – and the first half mile was used by cows twice daily on their way to the milking sheds. Anyone going out on the spree had to protect their party clothes by an old coat, and wear wellington boots until they reached the end of the drive, leave the travelling clothing in a ditch under an old zinc sheet, and then 'cloth up' on their return. There were hazards galore. On one dark night I cycled into a black horse and buckled my front wheel!

A popular diversion was a 'Shropshire Party', at which the men sat at one end of the supper table and the women at the other, instead of being arranged male and female alternatively. Presumably the men found farming more fascinating than the women, and could discuss matters in all seriousness, while the women preferred their inconsequential chatter about domestic economy more to their taste than the chaps were. Farmers are notorious for putting aside all distractions, female or otherwise, for a chat about fertilizers, the merits of this or that implement, or the happenings at last weeks' cattle market.

Can it be that women find men in general interesting all the time, whereas for men, this absorption runs in fits and starts? "Male arrogant rubbish!" I can hear my wife retort, yet there was a proprietress of a Chicago 'sporting house' who banned gambling in case the clients lost interest in her girls. As Robert Browning penned:

"Love is for man a thing apart,
'Tis a woman's whole existence".

Petrol rationing put somewhat of a curb on travelling for entertainment, but we could legitimately attend Farm Walks, usually organised by the War Agricultural Committees. Farmers put up little objection about visiting their professional colleagues, as they could carry with them some farming merchandise such as a bag of bran, or a horse collar, in the boot of their car. One farmer always took a calf, which must have eventually held the record for miles covered in Derbyshire.

Farm walks were held on a farm of a local enthusiast who had managed to squeeze extra milk out of his cows, or to grow a bushel or two of corn over the accepted standards. The key factors were the same as today's – management and timeliness, which I feel applies to industry generally, but agriculture particularly. Some years ago my own cows yielded 800 gallons of milk over a twelve month period. The cowman left and was replaced by another. The following year, the yield increased to 1,000 gallons with virtually the same diet, housing and herd. Only the operator was different.

Many people who ask "What is the daily yield of a cow?", might appreciate a useful rule. To calculate the daily yield at peak of lactation, divide its annual gallonage by two hundred. Thus a 900 gallon yield in a year indicates that the cow would give about 4½ gallons per day, soon after she has calved.

Attending farm sales has been a popular pastime of countryfolk for as long as I can remember. Unless one is there to buy, they are an enjoyable waste of time. Yet, the occasion can be tinged with sadness as someone's acquisitions, perhaps of a lifetime, are dragged out for all to see, and put under the auctioneer's hammer.

The rubbish is always sold first, to give the serious bidders time to arrive, then the more valuable stuff, and finally the livestock.

On occasions there might be surplus household furniture, which formerly offered a good buy for the discerning person with an eye for the bizarre or the antique. Nowadays there is so much interest in this kind of merchandise, that one must be very fortunate to get a bargain. The birds, early or otherwise, seem to outnumber the worms.

My personal best buy was a big, old farmhouse table, which holds pride of place in the farm office, and on which I am writing at this moment. The visit was originally made to buy a set of harrows, and I had made the successful bid. Meantime, my wife had gone a bit nutty over some disused stone pig troughs, which she thought would be ideal for growing her geraniums, and I sat on the table, waiting and ruminating about life in general. Suddenly the auctioneer appeared and sold the pine table under me. It was quite a bargain for eight shillings!

Broadcasting has provided the opportunity to travel widely, particularly in Britain, and I will always contend that for variety, interest and warm heartedness, our own country takes a lot of beating.

Take Yorkshire for instance. It is really a beautiful county, except for its industrial triangle, and if not, like Muhammed Ali, the prettiest, must surely be the grittiest, at least, its inhabitants appear the grittiest I have met.

Several years ago I took some time off the farm for a 'mini-tour' organised by the B.B.C. We started at Richmond with a performance of *Ask Me Another*, after which we took an incredible journey higher and higher up the Dales, to what must be one of the highest pubs in England, at Tan Hill.

We were back to nature with a vengeance, and, it seemed, also back in time. Even to enter the pub I was obliged to kick an old Swaledale ewe from the front doorstep.

In the quaint atmosphere of the Snug, with my pipe and pint, I reflected on the variation in scenery as one travels from the lush prosperity of the Vale of York, along the narrowing river bank to the wild moors, where farmhouses stood in isolation and sheep were dotted here and there in every direction. No wonder the Brontes were such prolific writers. There was little else for them to do.

Being a farmer I found it comparatively simple to judge the prosperity of each district as we travelled through, by the age of the farm machinery. What the farmer on the better land bought ten years ago as brightly painted red, blue, green or yellow monsters, had been relegated to farms further up the valleys. Now, rather tatty ageing machines (without cabs), they were helping the smaller yeomen earn a dogged living, while their former place of glory had been taken by even bigger, even brighter monsters.

Even in this enlightened age we have a pathetic ignorance of 'how the other half lives'. We carry out our daily work and relax, secure in the thought that our wages are good and that we are inhabitants of a fine Welfare State. The country will make sure that no one will starve. Yet in some of the wilder areas of the country, small farmers can experience the extremes of privation.

The situation in Yorkshire was driven into my mind with a bang on the occasion I addressed a Farmers' Union meeting at Brough.

Many of the people present recalled their terrible experiences during the frightful winter of 1963. They suffered unbelievable hardships. We in Derbyshire had a rough time of it too, but our frozen water pipes and tractors which went on strike in the cold were mere child's play, compared to the agonies and heartbreak of those whose livelihood depended almost entirely on sheep.

One dour Yorkshireman told me simply, without any embroidery, how his sheep, cut off from food and shelter, formed a triangular formation rather like a Roman phalanx, for their own protection. As conditions worsened, the animal at the apex of the flock bore the brunt until it died, his place being taken by another. This went on in succession, until only half the flock remained when the weather softened.

This may surprise those who consider sheep to be unintelligent, boring creatures. This is not the case at all. Given individual attention, I find them quite bright animals, but in the mass they are less so. Not unlike human beings really!

As a rule, the sheep stock belong to a particular farm, and when a change of occupancy occurs, the sheep stay put. There is a definite advantage in sheep knowing their own hills. They automatically find the sheep paths, and have an instinctive sense of avoiding dangerous gullies, bogs, and other dangers. They can find the best pasture in the shortest time. Another surprising but most desirable talent is that when a sheep is properly 'hefted', that is acclimatised thoroughly, he knows the boundaries of his own farm.

One can see, therefore, that the loss of breeding stock on a hill farm is almost disastrous. Restocking with fresh sheep is similar to sacking a team of highly skilled farm workers, replacing them with a crowd of unknowns sent from the Labour Exchange, and expecting them to run the farm with the same efficiency.

Despite their hardships, the hill farmers are hospitable people, and I was somewhat loathe to leave them, on the continuance of our tour.

Back we went, over the knobbly Pennines, and past the windy moors to a more industrialised part of the county. Here, between patches of blackened fields, little colonies of stone houses and factories have sprung up over the years, as a permanent monument to the Textile trade. In this area I found it difficult to distinguish one town from another.

Northumberland, to me, is one of the most interesting counties, and again I can thank broadcasting for the introduction.

I accompanied a film unit to Kielder Forest, which is one of the Forestry Commision's largest undertakings, for a B.B.C. Farming programme.

As a subject, forestry is a bit static for a film, even a documentary, and the producer, John Bird took considerable pains to get some footage of wild life, especially deer. The cameraman also tried to obtain some shots of the blackcock (or black grouse), a fascinating bird. At that time of year, the male, resplendent in his courting finery, performs a solitary dance at first light, to impress the grey hen. Isn't it odd how, in the bird world, the male wears the finest feathers? Deer, which are on the increase, are tolerated in forest areas, where the damage they cause is only a fraction of the havoc they can create on farming land. It is necessary, though, to keep down their numbers by systematic culling.

During my stay in Northumberland, approved sharpshooters were allowed the job, and, in fact, two deerstalkers were there at the time. One was a Polish emigré, known affectionately as the Baron. His colleague's name escapes me, but I remember he worked for the Ethiopian Meat Company.

Each evening, myself, the film crew and these two would relax in the local Working Men's Club which was run by a splendid lady called Agnes. Whether drinking and deer stalking go hand in hand is a matter of conjecture, but these two could certainly sink 'em. A double whisky to the Baron was merely a damp glass. After one particuarly stiff night's indulgence, I woke up with a desperate need to go to the loo. Staggering about in the pitch dark, without any sense of direction, dulled by sleep and booze, I mistook the wardrobe for the bedroom door with disastrous results!

The Kielder Forest area impressed me to such an extent that I returned, in 1965, but this was a more leisurely trip. I was accompanied by Marie and three of the five children we had then. The evenings were not so boisterous, and as we were on holiday, the days not so busy, as on my previous trip.

We had decided to try a caravan tour, but thinking that our crowd of seven would be too much a burden for this new white monster to bear, we took the three eldest children, and left the other two with a kind aunt. The caravan was hired, with a view to buying if we liked the exercise. Up to then I had always looked upon caravans with a certain amount of distaste – nasty, cumbersome dinosaurs, which cluttered up narrow lanes, delaying the more considerate motorist.

However, on driving the ensemble I developed an entirely different attitude, a feeling of independence, and a certain indifference to the 'road hogs' impatiently trying to overtake.

Which brings me to the farmers' attitude to the motorists who descend upon the countryside on high days and holidays.

Apart from biological functions, one activity we all have in common is travel. In this respect, I naturally feel that farmers, and particularly their wives, are some of the country's most competent drivers. Not for them is the smooth trip to the office on a metalled surface, but somewhat of a 'Rallye Agriculturale' every time they make the most routine trip, even popping down to the village shop.

The farmer can look at motorists from two angles. He is part of the public who consumes the crops the farmer grows. He may feel flattered that his particular domain is considered so attractive that people flock from far and near for the experience he gets every day. I personally feel that city dwellers have a right to the fresh air and beauty of the countryside. Yet on the other hand, the peaceful environment, where he has chosen to make his crust, is subjected to an absolute invasion of countless motorised strangers.

Imagine his thoughts, sitting in a tractor stuck in a four mile traffic jam, along a lane which he almost considers his own property. "I'll never get this blasted swathturner into that hayfield. Here I'm working while this stream of eager beavers are out enjoying themselves. What's that lunatic hooting at? Doesn't he know that this tractor's not sprung and that the bales might fall off onto his newly polished car if I go any faster? Too bad he can't see round this load of fertiliser, after all I am using it to grow food for *their* bellies. They didn't care when I wanted to bring my heifers across the road last weekend. Not B - - - likely. Next time I'll move the animals before morning milking – while this lot are still lying in bed".

I agree with his last point. Driving on country roads at first light is the best time to get along. All one has for company are the carrion crows picking at the remains of various rodents knocked off by yesterday's traffic.

It's bad enough to endure these motorcades on the move, but what happens when a family saloon disgorges its contents on, around or actually in his fields.

I imagine that all farmers recall the family on safari, Ma and Pa with the Primus going and enthusiastically watching the kids – presumably Mick McManus and Jackie Pallo in their own minds, rolling about – the lot comfortably situated in his MOWING GRASS.

In such circumstances, the best action for a farmer to take is to politely enquire if there is anything he can do to increase their pleasure, that he hoped the view was to their liking. Tell them not to worry about the odd broken bottle, tin can, etc. which they may prefer to leave with him, and to explain gently that plastic bags are really not the best diet for milking cows. They do say that sarcasm is the lowest form of wit, but I have found this is the best form of education. It is usually effective. To rush baldheaded up to such a family shouting "What the £*%@† ! ! ! ‡§&% do you think you're playing at. Get out of my £*%@† ! ! ! ‡§&% field and take your £*%@† ! ! ! ‡§&% rubbish with you" more often is met with stares of blank apathy.

At least, in this type of incident one can see the culprits face to face instead of wondering who left the gate open, who smashed the fence down, and horror of horrors, who was the idiot who didn't know that his old bicycle in a crop of barley could not be processed by a combine harvester?

Then, of course, there is the inevitable souped-up or steamed up, courting brake. A charming ritual, admittedly, but not when it's parked in the very gateway that one wants to get through.

Farmers do not think like this all the time, nor are all the public as irresponsible as I have described, and it is easy to see how the motorists react when the wellington, so to speak, is on the other foot.

A motor cyclist friend once ran into a horse on a very dark night, causing hospital treatment (and not for the horse), and speculation as to who left the gate of the field open. Like humans, animals are prone to stray from time to time and it is touching to realise how concerned the travelling public are for their well being. Many times have I been disturbed during what I consider a well earned rest, by a public spirited citizen who informed me "Your beasts were on the road, Ted, but I opened the gate down the lane, and let them in". It seems churlish when I have to tell him later that they were not my animals at all, and he had mixed them with some other cattle with which I had every reason to prevent my own making even a passing acquaintance. Farmers are making great efforts to rid the national dairy herd of that cursed disease brucellosis, and I cannot emphasise too strongly the possible dangers of mixing one herd with another.

There have been many theories and arguments as to whether one's personality changes when at the steering wheel. Many people take the opinion of an amusing continental cartoon, which starts with a mild little man approaching his gentle family saloon. Once in the driving seat he turns into a horned demon with a built-in snarl, the car grows fangs, and together they chomp their fearsome way, leaving a trail of destruction behind them. I believe it happens to me, but to a lesser extent, especially if I have more than 75% of the family in the car on a journey of over 150 miles.

Our kids are no different to the majority, but the thought of being incarcerated with all six, and Marie, say to Southern Spain fills me with panic. Those who attempt such brave deeds and still retain their sanity have my full admiration.

My own pet annoyances on the roads, apart from danglers in other peoples' cars, are holdups and slow moving jobs like caravans, balers, or cows coming in for milking. "Surely that chap could rev it up a bit", I think, taking up the general error of regarding the journey as a tedious way to get from A to B, and not to be enjoyed for itself. There, you see, I suppose this indicates that behind the wheel I cease to become a farmer, and take on the attitudes of the motorist in general. In fact, once on a holiday in Wales I was ordered out of a field by an irate farmer "I'm trying to make a living here, get out you English B - - -". I nearly reported him to the Race Relations Board, look you ! !

One's lifetime impressions can disappear in a flash – like mine regarding caravans. If your own impressions of country life is that there's plenty to look at, but nothing to do, give it a whirl! You'll find your own attitude will change too.

16

Recharging the Batteries

ONLY a fool would deny the value of travel, and I am one of the last to believe that the advent of cheap package holidays has been anything but of great benefit to all. Except of course, to some of the sleepy Spanish villages now transformed into something resembling Miami Beach, a sleepy village itself at one time, I suppose. If Franco has brought anything at all to Spain, it is surely concrete.

Let's see the World by all means. It needn't "Costa Packet", but let's not be brainwashed away from Britain. Too many people consider it the done thing to travel to far-away places on their holidays, before they've even seen their own beautiful country.

This mass overseas exodus at holiday times has only happened comparatively recently. I live in a small village in the heart of the Midlands, and only ten years ago, one of our old ladies died without ever seeing the sea. In fact, before the war a trip to London was considered as much an event as a trip to Siberia would be today.

Up to 1939, hardly anyone had been abroad – not counting the trenches of Flanders and Northern France. There was a tiny minority of 'gentlemen' or 'gentlemen's gentlemen' who could afford to travel overseas, but those who did so went on a single ticket to one of the Dominions or the Empire and stayed there, making their way I presume.

It's hard to realise that in the days of my childhood a foreign face was rarely seen. The Indian tea man was a curiosity with his turban and case full of packets, and attracted a veritable train of staring kids, whilst enagaging in his door to door sales campaign.

How different today! Our children have the opportunity of exchange visits, and so many visitors arrive from the Continent as part of the normal scene, that one wonders how long it will be before we are completely merged in the mainstream of European life.

Not only Europeans, but hordes of tourists from other countries descend upon Britain each year. It is not surprising to me that the charm of our islands as a tourist centre is spreading, but what a pity that so many of our own countrymen have not realised what they are missing.

I'm not talking about the traditional holidays on the coast, where funny hats, whelk stalls, amusement arcades and jam-packed beaches seem to act as a magnet to an awful lot of the population, but a holiday in the countryside.

Surely, the purpose of taking a holiday is to 'get away from it all', to leave the frustrations and tensions of the daily grind far behind. Where better than in the peace and tranquility of rural Britain, can one recharge his bodily batteries?

Most families have access to a motor car, which offers endless possibilities. By selecting the right routes, the physical process of travelling can be as pleasant a part of the holiday as the arrival. Of course, should you desire to explore the dual carriageway into Blackpool, that is your choice – certainly not mine.

When one flies over the country, one realises just how much unspoilt countryside there is to appreciate. It is very hard for the urbanite to realise that over 85 per cent of Britain's 56.2 million acres is what we understand as countryside – farmland, rough grazing, parkland and woodlands, and the area occupied by the latter is almost as great as the area occupied by towns and cities.

There is an infinite variety of scenery, customs, architecture and people – and of course, weather. It is probably the uncertainty of our British climate that sends so many people fleeing overseas, but I feel that the sheer unpredictability of the weather is, in itself a virtue. We island dwellers are saved from long, boring spells of heat, dust, frost, monsoon, drought, and the other unpleasant phenomena of large continental land masses. We don't have to spray our beaches with insecticides (only detergents), or struggle to keep flies out of our bedrooms. The rain washes the land, keeps it green, and we have no worries as to whether it's safe to drink the water.

I do my share of grumbling about the climate, as do we all, especially when the combine harvester is sitting in the rain, sheeted up and immobile, or when all the strawberry pickers are sitting under a hedge in a thunderstorm. They say that a good farmer never wants rain anyway. There is another saying in some parts of the land that when you can see the far hills it is going to rain, and when you can't it *is* raining. Even then, what better perfume is there than the smell of the land, or city streets for that matter, when a shower has laid the dust of a dry spell.

Britain has a tremendous variety of scene too, from flinty Norfolk to craggy Galway (well, it's a change from the usual Land's End to John O' Groats). One thing I have personally noticed though, is where the land is flattest the churches are tallest.

Have you ever experienced the weird sensation of an electric brae? It is a stretch of road which gives you the impression of going downhill, when you are, in fact, going uphill. So when you stop the car it runs backwards. I know of two, one in the Isle of Man and one in Ayrshire.

I have been told that some people actually prefer to live by night and sleep by day, rather like Mole, in the 'Wind in the Willows'. Presumably they burn up their time and energy in some cavernous night spot, from which they emerge, blinking at the morning light, with the strange scent of fresh air in their nostrils. Perhaps I was kidded, and such people do not really exist. Surely everyone in his right mind cannot fail to delight in the joys of the open air, away from all the trials and tribulations of his normal day to day existence.

There is a pretty reasonable choice of accommodation in country areas too. If you decide on one of our beautiful country towns as a centre from which to explore the surrounding district, you will be pleasantly surprised at the service and comfort of the hotels. You could even stay at a holiday camp, some of which are situated inland. It is not my place to be critical of these establishments, because I know some people enjoy this type of holiday. Three of my children once stayed a week in one and enjoyed every minute – I stayed one hour and left them in charge of an old and trusted friend.

However, if you do not care to be regimented, if you like your feet under a good table, and you prefer a free and easy atmosphere at modest cost, you could fare little better than spend your holiday on a farm.

This does not mean you have to share farmyard chores. Many farms now, particularly in the accepted tourist areas of the countryside, offer the same service as a guest house. The advantages are that you are usually deep in the countryside, but without the necessity of living out of tins, or even doing the washing up – a chore that every woman I know declares she loves to forget on holiday. I have a friend, a very enthusiastic fellow, whose ambition is to travel the greatest distance possible, in the time he has available. After a camping tour of Yugoslavia, his wife moaned to me that she needed another holiday to get over it.

People from urban areas are usually knowledgeable about the industries in their own towns and cities. On a farm one can see another type of industry, where livings are earned with the aid of machinery just as in the town, but it is all so different, and also refreshing to see the country doing its stuff at close quarters. After all, only a few generations ago, most of us no doubt had country connections of our own.

Even I spent a part of my own holiday one year on a farm. I took the family to Galloway. A busman's holiday, you might say, but I didn't think so. I wanted to forget my own farm for a while, and it was in the safe hands of my two men, Ken and Merrick. However, this was not to be, because in Galloway we discovered the hill called Merrick, and close by – guess what? – Loch Ken.

I'm convinced that people who, for most of their lives, are denied the simple pleasures of the countryside, need some contact with the land. The conservationists are also of the same mind. Although food production is the main purpose of the rural areas, the amenities of its surroundings are becoming increasingly important.

In my opinion, the best way to maintain our green areas is to allow a viable form of farming to continue. In the least favoured parts – the areas where farming is difficult – the farmer can receive part of his income directly, or indirectly from the leisure enjoyed by visitors. I'm not suggesting that National Parks become *natural* parks, because of the various facilities desirable to accommodate visitors, but I do think farmers make good wardens.

In National Parks, rural industry is supposed to carry out its normal activities, but the teeming masses from cities and particularly drab areas have free access. There are one or two regulations, of course, such as avoidance of the more horrific types of agricultural architecture.

There are, of course, many farms, parks, or natural areas where a holiday maker would find his time fully occupied – and not only by the facilities provided by the countryman himself, but by the nature of our own country. What is better than a walk along leafy country lanes, up hills and down dales, through woodlands and across grasslands. One can just laze by the side of a rippling river, or play cowboys with the kids. And let's not forget the hub of country life - the local pub. There's nothing better than a pint or two and a ploughman's lunch after a spell in the fresh air. No juke boxes, no pin tables, no one-armed bandits as in the chrome and glass palaces into which most town pubs seem to have developed. Just peace, good ale, and good company.

Successful campaigning for a shorter working week has meant that more leisure time is available to more of the population. In America, I am told that certain industries are now working only 3½ days per week, and who knows, this admirable state of affairs may not be too long in coming in our own country. However, additional leisure time must be put to some purpose and not be used merely to kill time aimlessly. This is when people can get into mischief whilst not slogging away at their jobs. I'm not alone in this view. Even government departments have been formed in an attempt to solve the 'problem' of leisure – not only for people with time off from their jobs, but also for retired people who, I suspect, find financial restrictions prevent them putting their new found freedom to the best use.

In days of yore, only the privileged classes had time to spare, yet, free as they were of the economic necessity to take up some useful employment, they appear to have ben no less frustrated than their less affluent brethren. When one considers the activities and irresponsible behaviour of some of the 'young blades' of the past, is it cynical to presume that the increase in leisure today may lead to general decadence, unless it is approached sensibly?

Just as yesterday's luxuries become today's necessities, when any service or facility or privilege becomes freely available, it is taken somewhat for granted, and eventually becomes unappreciated. Take Public Libraries, for instance. When compulsory education and complete literacy became the order of the day, reformers thought that people would be clamouring to borrow books. Perhaps they were in those early days, but libraries seem to have become a sadly underrated facility nowadays.

Public Parks similarly, are not used to their full advantage. In the past they were created with care for the benefit of the privileged few, but now are provided for the relaxation and enjoyment of everyone. One of the main objects in founding the original parks was the provision of sporting facilities, sport, incidentally, meaning the pursuit of something, deer, foxes, rabbits or even, I suppose, women, as the late A. G. Street used to insist, and not sport as we consider it today.

The majority of us who enjoy sport are members of the great legion of watchers, but I feel it is important that everyone should have some opporunity to exert one's bodily prowess in one way or another, even at low level. Physical exertion is an essential ingredient of life, whether in work, at play, or just walking. It doesn't require a psychologist to tell us that a couple of hour's hard digging or whatever, relieves depression and tension – it certainly does with me.

I think I share the view of most farmers about public parks, and gardening for that matter. We don't give them a lot of thought, but personally, some of the horticultural sections on the larger and possibly more formal parks fascinate me. I enjoy seeing the exotic plants cultivated in their thousands at Kew and similar establishments. Admittedly, I have a tendency to view these accomplishments with a peasant's eye of economic assessment, but why not? In the days of the old sailing ships, a stock of pepper, ginger or cloves was worth a fortune. Now that British eating habits have a greater tendency towards the Oriental, and the urban scene is becoming more widely influenced by the needs of peoples from far away places, perhaps spices may regain some of their medieval importance. In this country we can grow green peppers with a little care, aubergines are by no means a stranger to vegetable gardens, and who knows, I might even discover some profitable diversification. I can't really see a crop of bamboo shoots, bean sprouts or paw paws, though.

Travelling the world I have been able to see many exotic herbs, fruits, and plants of all descriptions growing in their natural location, and have even experienced the sheer joy of eating an orange freshly picked from a tree, but I have also seen a banana tree, complete with fruit in Darley Park, Derby, an establishment run by a Mr. Wells. This was quite an eye-opener, and not entirely because of its unexpected location. Up to then I had always thought the bananas grew upside down ! !

For more people to spend more time in the available space out of doors surely must be a safeguard against national insanity. There are masses of unspoiled countryside available, but the masses do not seem to realise that what was once the prerogative of a privileged few, is now available for their own use. There is a marvellous variety within reach of everyone – thousands of trees also, which are a particular asset. Apart from being things of beauty in their own right, they enable an area to accommodate more visitors without the appearance of crowding.

It is clear that more and more people actively pursuing their careers in the country are encouraging the flow of visitors into their territory, and I reckon that if there are any among you who still thinks of the countryman as some kind of peasant with straw in his ears, he will soon change his views. In a wet year, straw is too valuable to waste like that anyway, and they're talking about making it into paper too!

One of my favourite writers of essays is William Hazlitt. I once rented a house at Aldeburgh for a fortnight, too late in the year for the kids to enjoy a summer holiday, and being the last tenants of the season, we had the joy of the accumulated filth and debris left by all the previous occupants.

However one of the compensations was to find on a shelf, a book by the aforesaid Hazlitt.

One of his observations was as follows:

> "There is nothing good to be had in the country,
> or, if there is they will not let you have it."

Whether the masses get their hands on whatever is going depends on the attitude of the farmer. So far only a small minority have pursued the opportunity to accommodate the town visitor which sometimes causes their neighbours to react with suspicion or hostility – perhaps because they didn't think of the idea first.

Bitter experience has left a certain amount of prejudice against all persons passing through, or even visiting the countryside, with no obvious means of support. If farmers are to be the perfect hosts, let's hope that the rest will be perfect guests!

17

Seeing the World

MY EXPERIENCE of overseas travel did not really blossom until I was over thirty, and from then on, as a result of the T.V. business, I have visited many, many, foreign places.

I did, however, manage a couple of trips abroad before I started to work on my own account and became a slave to the old cows, and also a trip to Paris on my honeymoon.

The first time was with a young chap of my own age, called Geoff. We had planned to travel to the Ardennes, then to Paris and return. To finance the venture, we decided to travel light, and to take in our bags as much coffee as we could carry, since we had heard that it was in such short supply in Europe that we could sell it at an astronomical profit. As it transpired, the scheme was an absolute pain in the neck. The guilty feelings of going through the customs, hoping that the tins and packets stuffed in our socks and shirts would not be noticed, were bad enough, but in addition, Geoff. contracted some mystery ailment, under which he immediately took on the appearance of a Dalmatian dog. With his strength thus sapped, plus the effects of running his motorcycle into ditch during the previous week, he claimed immunity from carrying the baggage, so I humped the lot.

Finally, when we found what we thought would be a suitable cafe to trade our valuable commodity, and behind drawn curtains produced our merchandise – disaster ! We found that half our load wasn't coffee at all, but cocoa. We just about recovered our original capital, with a free cognac thrown in, and this terminated my brief career as a carpetbagger.

It would have been more profitable to have brought the few

pounds invested in our contraband as notes, for at that time it was possible to obtain more than the official rate of exchange from suspicious looking touts on street corners and dark places. These also specialised in postcards (clean and filthy), by day and the usual diversions offered by any large city by night.

The following year I again spent my week's holiday in Paris, with two other friends. Being typical mugs, we agreed to pay through the nose to one of these unsavoury characters, for the privilege of watching 'a spectacle'. It turned out to be a gymnastic display of two women leaping about a bed in a room on the top floor of a building. Just as we were leaving, and descending a steep circular staircase, the lights went out. One of my friends, thinking no doubt that we were about to be robbed gave a great push, and we all ended up in a heap at the bottom of the stairs. When we returned to our lodgings he was sick over the carpet, revulsed either by the show, the booze or the food he had scoffed – we were never sure which.

Paris seemed rather a jinx where I was concerned, because my third trip, on honeymoon, coincided with my first and second quarrel with my wife. I don't suppose anyone can fall out with his wife before they are married, can they?

In this case, the first argument was caused by my suitcase, which was a very old one, left to me by my grandfather. It wasn't very smart, but I never dreamed Marie would be so offended because I had securely fastened it with baling string—to prevent fallout, of course.

After three or four days of the holiday, my new wife reacted to my everlasting desire to visit old tombs, palaces and the like, by suddenly rushing from the top deck of the bus without warning, and to my horror, momentarily vanished into the crowd.

As I tried to restrain her, a large Frenchman tried to restrain *me*. Brushing him aside, I eventually recovered my spouse and discovered the reason for her dramatic evaporation from the bus. She wanted to see the smart shops on the Faubourg St. Honoré or at least the Galleries Lafayette – a fact, which being somewhat thick, I hadn't realised.

It's dangerous for an outsider to interfere in a matrimonial dispute. I learned this many years ago at Nottingham Goose Fair, where just outside the boxing booth I saw a man belabouring a woman. Thinking the time appropriate for a Boy Scout act, I, like a fool and not an angel, rushed in, only to be assailed in turn by the wife, for my trouble.

This third Paris visit was the last trip for the existing passport, and it was as a result of meeting Hilary Philip, producer of the B.B.C. 'Farming', that necessitated my applying for a new one.

He invited me to travel to Denmark in 1959, in order to make some films on Danish Agriculture. I remember 1959 was a very dry summer. This magnum opus was produced by Derek Smith, and he and I flew straight to Copenhagen, whilst the film unit, consisting of two cameramen and a sound recorder travelled by boat to Esbjerg. Derek and I met various people who arranged the programme for the film shooting, worked our way back across the country to meet the film team at Esbjerg, and then made the films.

I didn't know quite what to expect of Denmark but I had always wondered why the capital, Copenhagen, was right at the extreme east of the country. It was explained to me that in the olden days, parts of Southern Sweden, a place called Skone, were part of Denmark, and although one thinks of the Sandinavians as being perpetually at peace with each other, not far back into history, this was the theatre for savage wars between all countries involved, particularly between Denmark and Sweden.

It is a very flat country, but still has a charm of its own. A typical Danish farm had about a dozen cows housed in a big shed, with pigs at the other end. The entire shed had an insulating roof, because the fodder was stored up there.

On a larger farm, in Eastern Jutland, there were 150 bulls tied up by the neck in one shed, having been reared for beef. There seems no reason now to castrate male steers for beef production, as bulls grow quicker and better, and are even claimed to have more flavour if they are not castrated. This practice is now being advocated in many more areas of the World.

The Danes are very keen on intensive grazing. Where the cows are not tethered in traditional European style, neat paddocks were eaten off in rotation, by them.

One of their striking features is the widespread use of electric fences. I noticed they were connected up to the mains, and as soon as I returned home I obtained one of the transformer units, and it worked well for some fifteen years.

Electric fences took over from barbed wire. It is recorded that the first time barbed wire was discovered to have defensive qualities for military purposes, was when the Prussians invaded Schleswig-Holstein in the nineteenth century, when Bismarck started the first of the wars culminating in the establishment of Imperial Germany in 1871. The others were against Austria and France.

What particularly amused me in Copenhagen, were the rows and rows of Englishmen in the bookshops and libraries, engrossed in the unexpurgated works of Henry Miller and James Joyce, and thumbing through the uncensored picture magazines. Mind you, I was astonished at some of these publications myself. This was before Britain relaxed her censorship standards, and probably these 'sexy' books wouldn't warrant a second glance today in a Soho bookshop.

Not only are the Danes very keen bridge builders, with some fine bridges connecting the various islands of their country, they are also intrepid flag wavers. I've never seen a bigger display of flags in every possible position, and on every possible occasion. The Danish flag is called a dannebrog and is the oldest in Europe – just something I learned from one of the quiz shows I was on!

At the time, Denmark was relying as we were on exports, and its weakness was the predominance of agriculture. The two biggest customers were British breakfast – the old bacon and eggs, and the German mittagessen, for which the pork was favourite. Looking back, it appeared that the British breakfast was considered more important than other continental meals, because when Britain declined to join the Common Market the Danes followed suit, but now has re-aligned herself with the nine, the same as her best customer – U.K.

Because the Danes stayed in the same camp as the British, some of the German market previously held has, by 1973, become lost to them, but knowing the Danes and their ability to organise, it will not be too long before they recoup most of their losses.

The main difference between Danish and British marketing, is that the Danish trade is largely overseas, and the fact that they organise themselves into co-operatives. Butter, Danish bacon and eggs, all have to be channelled through some sort of processing plant or packing station, which means that a tighter system of grading can be imposed, and anything not up to standard, which admittedly is rare, is not exported at all. In Britain, most of our produce is consumed at home, therefore the reason that Danish bacon has such a high reputation in Britain is because they send only the best. In Britain all grades are consumed, although the best is probably as good as the Danish.

During our journeys we crossed the island of Funen, the capital of which, Odense, was the home of Hans Christian Andersen, and found ourselves in Eastern Jutland.

The Danes, like the Dutch, appear to make the best out of every bit of soil, and even reclaim where possible, which is encouraging and probably a very long-sighted investment. Some of the poor sandy lands in Eastern Jutland would, if in England be used solely as golf-courses. If one shuts one's eyes in this part of the country and listens to the dialect, and then does the same in Norfolk or the East Riding, one can detect a distinct likeness.

I'm always able to interest myself in bits of history in foreign places. An instance of this occurred in Denmark, where we stayed at an old castle called Dragsholm.

On arrival, late on a moonlit summer evening, it reminded me of what I imagined Count Dracula's pad in Transylvania must have been.

As we passed under the Barbican into the forbidding quadrangle, a huge dog, (obviously a Great Dane), ran out, baying, its the only word for it, at us or the moon or both.

It turned out that the Earl of Bothwell, who was married to Mary Stuart, Queen of Scots, escaped from his enemies to Norway, where he met some of the relatives of a former girl friend, and also some creditors.

Probably considered to be a useful pawn in the political game of time, he was eventually imprisoned at Dragsholm, by the King of Denmark, who was also King of Norway in the 16th century.

I have no idea how many years poor Bothwell spent in the dungeon under my four-poster bed, but *he* must have given the place its spooky atmosphere.

As far as my getting around was concerned, the best bit of luck was when the phone rang one day, and Dick Maddocks from Birmingham invited me to take a place on the panel of *Treble Chance*. This was the Radio quiz programme on general knowledge and music, produced by Michael Tuke-Hastings.

At the time I was considering an offer from Angus Wright of Southern Television to do a regular series of programmes, but decided that the regular chore of travelling between Ticknall and Southampton was too much. This was ironic, because *Treble Chance* led on to *Forces Chance*, which, in terms of mileage per minute transmission time, must hold the record – at least, as far as I was concerned.

Treble Chance consisted of a regular team, plus question master, which travelled to different places (outside broadcasts by definition), and competed against local teams. Sometimes the teams represented Universities, seaside towns, local radio stations, etc. The programme finally changed in style to include Patrick Moore and myself as resident captains of local teams. In Patrick's first programme, the timetable had to be altered to enable him to contribute to the final splashdown programme on television, of the Apollo moon shot.

In *Forces Chance*, the teams we played were selected from military, naval or R.A.F. units. The team captain was Neil Durden-Smith and the question master, Alun Williams. Our lady member was a girl called Pamela Donald, who left to marry a Sherwood Forester captain whom she had met in Abu Dubai, and as far as I know, she is still in the desert. Her place was taken by Nan Winton.

My first trip for the programme was to Singapore, Malaysia and Gan.

Gan is a small island, only slightly larger than my own farm, situated at the southernmost end of the Maldivian chain in the Indian Ocean. A staging post for the Eastabout route to the Far East, it is the nearest thing to the Palmolive Soap commercial I'll ever reach.

Located less than 40 minutes south of the Equator, one can see the Plough and Southern Cross in the night sky, at the same time.

What a place for skin diving too! Not being used to such equipment, though, I had a very nasty moment, when I got water in the mask and started to choke. Meantime I was lacerating my knees on the sharp coral, being unable to stand up in the strong current as I had lent my flippers to a friend. After this, however, I improved sufficiently to enable me to appreciate the wonderful life of fish, turtles, porpoises and vegetation that flourished on and off the reef. Just like Captain Nemo in Jules Verne's *20,000 Leagues Under the Sea.* A much idler way to see the wonders in the crystal water was to borrow a glass bottomed boat.

Apart from the R.A.F. chaps, the only people on Gan are local Maldivians, who, each day, come from other islands in the atoll to work. The only female resident is a W.R.V.S. lady, and there is little other bird life either, although the nightly screeching of huge fruit bats might cause the casual visitor to think otherwise.

After stopping at Gan on the return trip, our party visited the barren island of Masirah, which lies off the southern coast of the Arabian Peninsular.

Two things stand out in my mind. Firstly the village built almost entirely of forty-gallon oil drums where most of the inhabitants lived, and secondly, the turtles which come ashore to lay their eggs on the sand about two o'clock in the morning.

It is a rare and rather wonderful sight when these giant creatures, so slow on land yet so agile at sea, dig themselves in and lay vast clutches of eggs on the surf beach. Being interested in wildlife, I found this place absolutely fascinating. Besides the giant turtles, there were gazelles, camels, nightjars, eagles and lobsters to excite my interest.

On the last morning before departure, I arose early and decided to look for some of the turtles' eggs I had seen being layed the previous night. On a borrowed bicycle I set out on the four mile journey across the arid landscape, passed behind the jebel, a rocky line of low hills, and arrived at the sea shore. Alas, at this point my chain came off and having no tool kit I was faced with the prospect of either spending the time available to put it on again, or abandoning the machine and sprinting back to the departure point. Luckily, I managed the former, arriving in the nick of time, heavily perspiring and with hands covered in swarf.

We went to Singapore – lion city of the East, and also into Malacca which was full of old Portuguese inscriptions and very interesting.

Whilst in Singapore I went sailing with Chris Foulke, a relation of mine who was in the Army at that time, and I also had a great thrill from diving deep into the sea and bringing up a piece of coral. I came back richer for the experience, plus two crocodile handbags, some Thai silk and a bunch of orchids purchased in Changi village before the take off.

Singapore, I discovered, is about 80% Chinese, whose industrial qualities seem unaffected by the equatorial climate.

Certainly it is a gourmet's delight. When I was there in 1969, one could eat from the gutter a sumptuous meal for five shillings, and even in 'Fatties' restaurant in Albert Street – the most well known – it was only a little more costly.

In nearby Bugis Street, the main attraction was the unofficial parade of transvestites who were, in fact, Chinese youths immaculately dressed to look like ravishing women. Their hormone displacement must have been such that they appeared (although I didn't get too close) to have breasts, and the only way that one could tell that they were not real females was to look carefully at their Adam's Apples.

Despite all this performance, apparently they resented being mistaken for real women, as one drunken sailor discovered to his cost when one of them turned really rough. "This could never happen in Ticknall", I thought, and whilst I was ruminating on this paradox, a precocious Chinese girl, about eight years old, challenged, and beat me at noughts and crosses at the cafe table where I was sitting – then demanded a dollar for winning!

Across the Causeway, which separates Singapore from Western Malaysia, I was fascinated to see the rubber and palm oil plantations, and thanks to the kindness of my friend Glyn Williams, who was a farmer's son from the Denbigh Moors in North Wales, residing in Singapore, I was able to get off the beaten track and see how the ordinary people lived. After all, it is the people with dirty fingernails who create the real wealth of any country, and I was determined to try my hand at planting rice in one of the paddy fields, which I did.

Among the places I saw was a privately owned Chinese farm with fields of pineapples, sweet potatoes, patches of pepper plants, and the inevitable carp pond, an Indian village with Hindu Temple, and another similar settlement inhabited by friendly Malays.

Friendly? well all but one little boy with a withered leg, whose spikey black hair and round face reminded me immediately of my son Tommy at home. When the other children gathered round the car he hid under one of the wooden houses. He needed a piece of wood to get along with, using this as a crutch. I was filled with thoughts of adopting him, but came back to reality after a minute's reflection on the impracticality of removing such a small human from his native land, climate, language and family, to a cool damp island off the west coast of Europe. Instead we left a few dollars with one of the village men (without affronting his pride, I think), to buy the child a proper crutch.

Singapore is a very clean place, due to the energetic efforts of Lee Kwan Yew and his government. A man was even fined for flicking his cigarette ash onto the pavement. At least, that is what the papers said, and I believed it.

The following year, *Forces Chance* took us to Germany, and I took a look at the Berlin wall. I'd never been to Germany up till then except down the Rhine, and up the Moselle valley which was a little trip I had with Marie some years earlier.

We then went to Gibraltar, Cyprus and Malta the next year. Cyprus is a lovely island. I couldn't help noticing the division between the Turks and the Greeks. The Turks have the old part of Famagusta with Othello's tower in it, and also an area between Nicosia and Kyrenia, which we could drive through, but no Greek entered these Turkish areas.

The most interesting part of Cyprus, I found, was Pathos where, on my day off, I paid a visit to the place where Aphrodite is supposed to have arisen from the sea. Cyprus is steeped in mythology and ancient history. Wherever one sticks in a spade, there is sure to be some relic or remains to be discovered.

It is also a mine of interest for anyone with a connection with agriculture. The best oranges and lemons I have ever seen, grow on a large estate to the South of the Island, and it was with a tinge of envy that I learnt that the Cypriots harvest their barley in *April.*

Here again, there is a marked contrast in the styles and levels of agriculture. One can see farmers operating giant combine harvesters, then, not much farther down the road, come across herds of flop-eared sheep and goats, just as depicted in Biblical times – shades of Laurence Durrell's 'Bitter Lemons', in fact.

I tried to find a strawberry farm, in case I could export them some plants, but was told that their stocks originated in California!

One particularly interesting journey was even further afield than Singapore. We flew straight to Hong Kong.

It wasn't exactly straight, of course, because the R.A.F. air support command can never fly direct. They have to go over only the countries which are friendly towards Britain, which means they have to take a slightly longer route than a civil airline. On our route we flew over Italy, Greece, Turkey and Persia. Each time I have flown over Persia, I have thought what a fantastic place Teheran is; one day, perhaps I'll see it on the ground.

From Turkey we flew on to Gan, stopped once more at Singapore and finally arrived at Hong Kong.

I was amazed to discover that civil airlines crossed over Vietnam while the war was in progress, as if nothing was amiss. From the flight deck one could see the Mekong river and the smoke trails of the B52's. It was as if we were in a different world, six miles up, and one could ignore everything occurring on the ground.

We arrived exhausted on a Monday afternoon, and I ordered a new suit.

On the Tuesday we were entertained at Gun Club Barracks, the headquarters of the Black Watch. Whilst there, I met James Watt and Captain Pirata, who I previously met in Gibraltar, and also tried playing the bagpipes. It is most difficult to co-ordinate the pumping movement of your arms and also your blowing.

On Wednesday the tailor arrived with my suit and two pairs of trousers – how's that for service?

If there is one thing that the Chinese make well, it is mouth organs. I just couldn't resist buying one, and also two coolie hats, one for a male and one for a female. Quite unsuitable for Ticknall, however.

From South Hong Kong Island, I saw the hill where the First Canadian Division was wiped out in 1941 and also the house, formerly a nurses' home, where a terrible massacre occurred in the same year. Full of rather morose thoughts, I had a drink with Tommy Trinder who was doing Cabaret at the Mandarin Hotel, and who better to cheer up one?

On our various air trips to different R.A.F. camps, we passed over grandeose scenery, not unlike the Mawddach estuary in Merioneth. On one flight we experienced the sad sight of the great Queen Elizabeth, flat on her side in the water, due to the fire which destroyed her.

We went up to the New Territories, right up to the Red China border and gazed across into that mighty land. Unlike the Berlin wall there is no obvious barrier, in fact, we were told that farmers come over the border to farm their fields on our side.

Forty miles away, across the estuary of the Pearl River, is the tiny Portuguese colony of Macau, with a fascinating, seedy Grahame Greene kind of atmosphere. A daily hydroplane service makes a day trip an interesting possibility from Hong Kong.

It is overlooked by the Chinese Hills and has accommodated Jesuits for over 400 years.

I had fish for lunch – there seemed to be more of that than anything else.

The largest edifice is a seven storey fun palace. On the ground floor were fruit machines, the first floor had roulette and black-jack, and there were shops on the second floor. Higher up one could get a little consolation if one had been unlucky at cards, I think ! !

Later, walking round the harbour, I saw the hordes of Chinese people who live on house boats, and noticed that there were no seagulls, only kites. Perhaps it is because seagulls do not thrive in the tropics, but I think the real reason is that the diligent Chinese throw too little away for them to live on.

Hong Kong is a British Crown Colony. The Governor is the boss together with an Executive and Legislative Council, which are both by appointment. It must be the only place in the world where the population is too busy making money to worry about electing their own representatives, let alone becoming independent from the United Kingdom.

By chance I met Dick Hughes, an Australian who writes for the 'Sunday Times', amongst other things. During our chat over lunch at the Eagles Nest he happened to remark how well George Brown had handled the situation during the Communist riots of 1967.

I was pleased to learn that my M.P. until 1970, had made such a good impression, as most people consider that George had a fairly rough ride as Foreign Secretary. The next time I saw him on an *Any Questions* programme, I mentioned his Hong Kong fan club, to show that his efforts had at least been appreciated.

At any rate it proves how we remember things parochial and all mistakes, yet tend to forget successes, particularly if they happen in far off places.

My travels in the Far East did make me realise a great factor that I had previously taken for granted – the Seasons, as we have them in Great Britain.

Take Singapore for example – green and lush – but oh, how it rains. You never know whether it is Easter, August or Christmas Eve. The climate is always the same; as one leaf falls another grows. It goes dark at the same time every night. No long evenings, no short dark, afternoons.

It is not only Robert Browning who thought that the Spring in our own country is the best anywhere.

I have already mentioned Germany, but only in passing. My travels have, in the main, been across the north German plain, Westphalia, and Lower Saxony, where the farms are very tidy and the asparagus fields flourish.

German architecture impressed me by its solidarity, especially the way that the average citizen's house is constructed. Nowhere did I see the unattractive housing estates that have become a feature of the British suburban landscape. Most of their dwellings have cellars and lofts, but I suspect that accommodation for more than one generation, or even more than one family, is the rule rather than the exception.

The contrast between The Federal Republic and the G.D.R. is still striking, but less so than on my first trip. One wonders what will be the eventual form of the Communist system in an advanced technological society. In the East, essentials like food and housing are cheap, but luxuries, including motor cars are either very expensive or unobtainable – or both.

It is difficult to see how any liberalisation can be effected without the collapse of the system. Can the flood gates be opened just a little, without letting *all* the water escape?

Perhaps I have given the reader some idea of my lucky breaks concerning travel, resulting from my accidental involvement with broadcasting. Space prevents me from describing in detail other places I have visited, including Malta, Gibraltar, Tangiers, Bruges, where I visited the College of Europe, and Luxembourg.

On my only trip to Spain, I stayed at a friend's house at Cadaques, memorable for me because it was close to the exotic home of Salvador Dali, at Port Legat, which features in many, if not all, of his paintings as part of the background. It's not every day that you see a cheetah walking into a pub on a chain, but some folk, I suppose, just have to be different!

Having looked down on Italy several times en route to the East, I thought it was time to see that country for myself at ground level. Marie and I have been on a couple of package holidays. One to Naples, Sorrento and Capri and a quick four night/five day trip to Rome. Unfortunately going to such a place is really only like an hors d'oeuvre, whetting one's appetite for more.

As far as holidays are concerned, I firmly believe that the company is the most important ingredient. The best holiday we ever had was at Skegness!

18

Across the Atlantic

IN THE autumn of 1965 I received a totally unexpected, but very acceptable invitation to visit the United States of America. The trip was originally instigated by an American pharmaceutical company, Eli Lilley, whose headquarters were in Indianapolis. Their affiliated company in England specialised in drugs for use in the pig and poultry industry, and had arranged for about twenty-five leading lights in these occupations to take a three week tour of the Eastern Seaboard and the Mid West, to see methods adopted by their American counterparts.

As I had no pigs and only half a dozen hens, I was more than a little surprised to be invited.

In the 1950s and up to the mid 60s, the English climate seemed to consist of wetter summers than of latter years. Of course there were exceptions like 1954 and 1959, which, incidentally, corresponded roughly to vintage years for wine making. Not that I'm much of a connoisseur of wine, but I have noticed that if we have a dry harvest time in England, it suits the wine experts.

Maybe things are changing now, for during the last few years it seems that no proper summer weather arrives until the days begin to shorten. However, in 1965 the harvest was very late. My own hopes of finishing all the corn before I left Great Britain were obviously doomed to be unfulfilled, as Bert Ford and his mates persisted in sending over depressions from mid-Atlantic.

The season was so late, in fact, that we started lifting potatoes before we had finished the corn, because good weather is not so essential for potato harvesting as it is for cereals.

The trip was due to begin on October 11th, and thirty acres of sodden laid barley still needed to be salvaged.

Although so late, this was not a personal record. One year, I think it was 1958, I finished combining on November 5th, a very odd prelude to our Bonfire Party!

Nowadays, crossing the Atlantic is so commonplace that many friends of mine regard it with no more feeling than they would a trip to London in the old days, but to me, the flight was a big thrill. Besides having a rare look at the tip of Greenland – I always get a kick from looking out of the window at the land, marvelling at the cleverness of the map makers – I was delighted to see blue skies and the portents of anti-cyclonic weather, which I hoped would reach Ticknall in a couple of days.

I often wonder how far away is the day when some kind of visual communication will become readily available, to eliminate the rushing about of so many people, with the attendant fatigue that often goes unnoticed. For instance, some kind of apparatus incorporating a telephone plus T.V. set – all signals bounced off a satellite of course – would save an enormous amount of time and trouble.

Travelling westwards, if you fly at approximately 1,000 m.p.h. presumably you would stay for ever at the time at which you took off. At the Equator that would certainly be true. As the route to New York lies nearer the North Pole, it seemed that the day was endless, because the shorter distance between the lines of longitude gave me the impression that it was still roughly the same time of day when I arrived.

I have always been a mug for 'having a go'. Once, at a local fete in Leicestershire, I was persuaded to try the trampoline, and after a tremendous leap, fell with the agility of a dead donkey upside down on the back of my head, causing great mirth to the spectators, and the prospect of being another Quasimodo for the rest of my natural life.

Thus, with my body clock all in a jangle, I was still fool enough to accept an invitation from one of my American hosts to try a few bars in Indianapolis at nearly midnight, or six a.m. Derbyshire time.

At one particular night club cum bar there was a Twist competition in progress, and in response to some nut who kept saying "We have with us tonight folks, some gennulmen from England", I was persuaded to disport myself on the dance floor. The judge was either crosseyed or highly sympathetic, for after about fifteen minutes or so of this exercise, someone announced that I was the winner, the prize being a bottle of Scotch.

The whole event was so unimportant that I was amused to hear, in a letter from Marie, that it had appeared on the front page of the Derby Evening Telegraph. Such was the organisers' enthusiasm to get publicity for the tour on both sides of the Atlantic. Not the sort of reportage that goes with a serious study of agricultural practice, one would think.

The only other nightspot I tried was the Playboy Club in Chicago, – the original establishment of that benefactor to the human race, Hugh Heffner – but on this occasion I was so tired that I fell asleep and failed to notice anything good or bad about it, Bunnies included.

Our arrival was heralded by the first of many effusive welcomes. I found that Americans are seldom short of something to say. Although sometimes longwinded, they have a curious ability to find the most appropriate word, even if it has many syllables, and may even be a recent invention.

They also have a refreshing directness, illustrated, for example, by the way an American will introduce himself as he shakes hands with a new acquaintance, which is far better than hanging about, waiting to be introduced by someone else.

Perhaps this is a legacy from the old frontier days when it was essential for one man to know the identity of a stranger. In England, with our tradition of a largely static population, it is usually sufficient to take a man at face value if he is a friend of a third person, who could, I suppose, vouch for his probity. With a more mobile crowd of citizens, it is far more important to make a snap judgement in case the new chap might be a threat to one's security.

Another feature I noticed in all conversations I had with the natives, was their absorbing interest in their own ancestry, particularly if they had any British antecendents. British people are perhaps unaware that America is the melting pot of Europe, in spite of the fact that the original thirteen States involved in the War of Independance were wholly British in outlook, and largely so in ethnic composition.

I was able to pay a short visit to Virginia, which above the Mason Dixon Line, was still the most valiant supporter of the Confederacy in the American Civil War of 1861-65. It is a personal theory of mine that this war was fought between two factions of the population holding very similar ideals as the two sides in England, at the time of the struggle between the Cavaliers and the Parliamentarians. Certainly the New England States were peopled by the Dissenters, Puritans and the non-conformist element, while the Southern States had a good deal in common with the attitudes of the English aristocracy, which in the first place provided many of the settlers.

Of course, it would be stupid, if not arrogant, to make sweeping generalisations about a collection of two hundred individuals, just because they live in the same country, even one as vast as the U.S.A., nevertheless, they all seem to acquire that same American appearance after the first generation. One can imagine the efforts made by the authorities in the past to weld shiploads of immigrants into some sort of homogenous mass as they arrived, full of hope and aspirations, from the Old World. Perhaps that is why the Constitution and the flag – prominent in private houses wherever one goes – have such an important place in the hearts and minds of the average citizen.

During the tour, our party saw many fascinating agricultural enterprises, particularly in the Mid West states of Illinois, Indiana, Iowa and part of Ohio, where lies the 'Bread basket of the U.S.A.' The terrain was scrubbed flat by glaciers in the Ice Age, and has soil seven feet deep, in which grows maize and soya beans mainly. The animals are kept in huge feedlots in open air corrals, the cattle coming from the ranges in the west and south west. Another interesting feature is the large pig (hogs in the vernacular) units.

I was pleasantly surprised by the city of Chicago, gem of the Mid West and truly described as the premier marketplace of the agricultural capital of the world. Standing on the shores of Lake Michigan, it seemed a most attractive place. I looked round for gangsters but failed to spot any.

The Corn Exchange was remarkable. We arrived a few minutes before ten a.m. and gazed down from the gallery on to the floor of the building, where small groups of men were standing around awaiting the bell to signal the commencement of the day's dealing in corn, oats, wheat and beans, each in a different 'pit'. Suddenly the action started and the whole scene changed into a crowd of apparent lunatics, shouting and yelling and waving bits of paper. The electronic indicator board spraing into activitiy, as hundreds of deals were carried out in these commodities and also in corn futures, which is, in fact, gambling on the future price of any crop.

It was a sobering thought that what was being effected before our eyes, chaotic as it appeared, would have a direct bearing on the lives of all in the western world.

For the next few days each member of the party had the privilege of staying in the home of an American farmer, and I was entertained at the home of Vernon Miller and his family.

Vernon was a tenant, but unlike his counterpart in England, he did not pay a previously agreed rent. He was more of a share cropper, halving the cost of seed and fertilizer with his landlord, and providing the labour and machinery in return for use of the farm.

The farm was quite small, about 250 acres, and part of a section, as the land was laid out in squares. He grew only maize and soya beans, and as he worked the holding himself, I wondered why he owned three tractors. He told me the labour force was augmented by casuals at peak periods. Many Americans have two jobs, and even Vernon was part time lineman for a telephone company.

His grandfather was originally from Germany, and their religion was Mennonite, basically puritan which eschewed alcohol, tobacco and military service. Much of Sunday was spent in church, which I gather is typical of rural Illinois. Vernon had travelled to Chicago several times, and for holidays in the Rockies, but never to the seaside. What he possessed he had certainly earned!

After leaving the Millers' farm, I stayed at several small towns, before spending a weekend in Washington and a couple of days in New York City – not, I would say, typical of the U.S.A.

The small towns had much in common in their straight streets, and always the four illuminated signs, indicating the important places in the needs of the community – Gas, Drugs, Bar and Hamburgers.

We passed large tracts of uncropped land, called the Land Bank, the occupiers of which were being paid to leave it unplanted, provided the weeds were mown. This was just one example of the restrictive practices needed to put stability in the price of farm produce in a land of fantastic potential.

In Washington we did the usual rounds of the sights, the Lincoln and Jefferson Memorials, the White House, the Smithsonian Institute, the Arlington Cemetery and the Capitol. By comparison, New York City seemed the least desirable place to live. Set in the middle of Manhattan Island, it is hard to realise that one is on an island at all.

From the top of the Empire State Building, one can see the meaning of vertical integration. Being naturally curious, I asked a fellow visitor to identify a certain landmark in the distance, only to find that he was a member of the Leningrad Symphony Orchestra on tour, and considerably more in the dark than myself.

During one afternoon, spent in the United Nations building, I wondered if the organisation would have been more effective had it been constructed in a more neutral part of the world like Switzerland or India. However, there it was, surrounded by every flag of the member states, and I suppose functioning at least as a speech factory, and at most as a real agency for human compatability. One could listen to the debates by earphone in one of several languages, English, French, Russian, Spanish and Portugese. Having visited both Houses of Parliament in Westminster, I found the U.N. debates only slightly less boring, but I would guess that the cut and thrust and witty ripostes that occur in the House of Commons on supplementary questions, would be a far sharper form of entertainment than any equivalent in the United Nations building.

The entire U.S.A. tour brought home to me the fact that many of the influences affecting us in Britain have made the quick flip eastwards across the Atlantic. Even names! I had always believed, in my ignorance, that the Light Sussex hen came from Sussex and that the Rhode Island Red came from Rhodes, but no! The former is a native of Sussex County, Delaware, and the latter from the smallest state in the Union.

Television, in colour and at breakfast, now in practice in Britain, was commonplace. Even the commercials as we know them have a distinct American birth certificate, by style at any rate.

Super and hyper markets, then not quite so numerous in England were the normal places to buy the necessities of life, including milk, which was not meticulously delivered to the doorstep as it is at home. I would imagine that the British housewife, and of course, thousands of milkmen, would pull out all the stops to prevent this facet of American life catching on in our own country.

Their attitude to commerce and industry, generally tougher towards its executives, it appears, seems to have been adopted here too.

However, Britain could certainly take their lead where pedestrian discipline is concerned. No one walks across the road against a red light in their cities. Another factor of the American way of life which impressed me is their cleanliness. For some time I was under the impression that the aroma which hung around Indianapolis was corn oil, but discovered that it was a kind of floor cleaning fluid, instead.

In a society where the individual is supreme, I noticed a definite aversion to anything which smacked of socialism, to the man in the street. I was amazed how unacceptable was the concept of a National Health Service, and couldn't help noticing the sections of the big cities that had become ghettos for minority groups like Negroes and Puerto Ricans.

One vivid exception to the west to east influence, came over loud and clear whilst I was there. It was the Beatles ! !

My stay in the U.S.A. was all too short, for what can a man see in such a vast country in such a limited time?

By contrast to the outward trip, the homeward flight seemed not an extended day, but rather a condensed night.

I arrived home via Broadcasting House, where I did some wild track recording for a record programme. The corn was all in and most of the potatoes, except for four acres of King Edwards.

It was still so wet in England that these stayed in the ground until the following Spring, by which time the frost had rendered them useless. Another first experience !

19

Reflecting on Travel

ONE of the big advantages of living near to Derby is that it's half way to anywhere in the British Isles. One can catch a train to most parts, and not be faced with the prospect of changing at too many stations. In the days when those magnificent big engines *steamed* their majestic way up and down the country, Derby Station was an acquisitive hunting ground for hundreds of train spotters. Every day dozens of express locomotives, would stop there, or thunder their way through, proudly bearing such names of 'The Black Watch', 'Zululand', 'Tasmania'. Diesel engines are certainly quieter, and probably more efficient, but I doubt whether they have the appeal of the old steam locomotives to the general public. Railwaymen say the old steam locos even had a personality of their own, and who am I to disagree with them?

I can remember clear frosty evenings, sitting up with a cow expected to calve, and felt some kind of companionship with the goods trains which I could hear chuffing and clanking away in the distance.

The first farm on which I worked, Shaws Park, Stanford-on-Soar, was adjacent to the old Central Railway Line from Sheffield Victoria to London Marylebone. The farm was occupied by Mrs. Brown, and run by her son Charlie and his wife Dorothy, who worked harder than any woman I have ever seen. The staff consisted of an old twisted up waggoner, a concientious objector and myself. There was one train which left a tunnel as regular as clockwork at 12.30 every day, and we used to knock off for lunch as soon as we heard its whistle.

Naturally farmers had mixed feelings about the old steam trains, which were certainly a fire hazard. After the binder had made the sheaves at Shaws Park, all had to be carried about 30 yards from the railway line to minimise the risk.

On one occasion a spark from a passing engine set our neighbour's hedge alight. The farmer and his men rushed up to the hedge from his fields, the railwaymen rushed down from the tracks, but the farmer wanted the fire to reach an old stack of poor hay – if the beasts couldn't consume it the fire could! However, the railwaymen won and put the fire out.

My first and last ride on the route, 25 years later, was just before passenger traffic finally ceased. As I passed the old fields I remembered the day when a wild heifer got on the line and stopped a train.

I enjoy travelling by train, and if there is a good, or even reasonable service to my destination, I prefer to let British Rail take me. Motoring can be so tiring and dull if one is driving to an appointment, whilst rail trips I find restful, relaxing and in a lot of cases quite eventful. I once heard of a woman sitting in a railway compartment with her child, when a rough looking man entered. Feeling a little apprehensive of his appearance she said "I think I ought to warn you that my little girl has measles". "That's all right", replied the man, "I was thinking of committing suicide in the next tunnel".

The supposed reticence of the British to speak to their fellow travellers is grossly exaggerated, in my experience. I suppose it originally arose through people sitting in trains thinking to themselves "I wonder if that fellow over there would like a chat", instead of opening a conversation themselves.

When Marie and I were in Dusseldorf we became involved in a conversation with a waiter at a restaurant, who complained, as he served the boeuf Stroganoff, that when he worked in Britain, he had journeyed from Euston to Blackpool without a word of conversation between himself and his fellow passengers. "That couldn't happen anywhere else in Europe" he maintained. At that moment one of his colleagues was being harangued by two pre-1914 vintage German ladies in such a stentorian manner, that their dialogue could have been heard by everyone within a radius of a hundred metres, I thought.

His remarks made me wonder are the British shy, or is it that the majority have not the same need of conversation as, say, the French?

Yet I have found this not to be the case, and I think a train journey is an excellent occasion to broaden one's spectrum of society. Most people are amicable, and it is remarkable how even the most stern and apparently unapproachable faces relax once they are engaged in conversation.

On one trip, from Derby to London, I travelled as far as Leicester with an expert on South African oranges and lemons. I learnt that lemons are never out of season, among other things, and that they are imported to Britain all the year round from different parts of the globe.

His seat was occupied at Leicester by a very presentable young woman, who was a croupier in a gambling club. Incidentally, the proprietors euphemistically preferred to call it a 'gaming club'.

Presumably they thought that gaming sounds more sporting than gambling, and I suppose they're right. Gaming seems to paint a picture of lace shirts, Old English game cocks and snuff boxes, rather than a game of poker in a hotel bedroom, or picking the winners in a back street betting shop. Whatever they call it, it seems to me that the mugs always lose.

Admittedly, opening up converastions with people on trains does involve a certain element of risk, but the benefits in most cases make it worth taking the chance. One can, of course, be committed to a hundred or so miles of monologue by a crashing bore, who will even persist after one has signalled defeat by burying one's head in last Sunday's newspaper (quality of course), in an effort to catch up on the news.

Your companion might be a mutterer, barely audible in normal circumstances, but who will reach the 'punchline' of his story as the train is passing through a tunnel, or when a passing express roars by. His every word is drowned, consequently your expression of complete withdrawal is taken as an inability to comprehend what he considers to be a matter of absolute clarity.

There is, of course, a risk of being completely snubbed by the person to whom you have addressed your opening gambit, and this can be embarrassing. I can do little better, personally, than to take a leaf from Stephen Potter's book 'Lifemanship', and to adopt a ploy which could put the offender above me in the scale of discomfiture. For example, if your overtures are met with a diffident glance and a deliberate return to the Financial Times, why not whip out a crossword, preferably from The Times, Telegraph or

Guardian, and complete it immediately. The harder and more complicated the crossword the better, particularly if it is a well known one. There's no need to even bother about clues. Just put down any old letters indiscriminately. On its completion, fold up the newspaper and settle back with a rather superior expression. It sounds a pointless exercise I agree, but one certain to get some reaction from your fellow traveller. A word of warning. Never lend him your newspaper after such an operation!

It is becoming increasingly possible that one could have a most undesirable travelling companion or companions, particularly on Saturday. I have never had the misfortune of actually seeing a vandal at work, but I have travelled in carriages which must have been occupied by someone of this ilk, shortly before. There is some pretty antisocial behaviour that takes place before all and sundry, however. Habits like filling the compartment with cake crumbs or the remains of a bag of crisps, and slopping lemonade onto the floor, the seats and everywhere that lemonade can be slopped, or using the opposite seat as a footrest, whatever filthy condition the travellers' shoes may be in at the time. The carriage cleaners employed by British Rail certainly have *my* sympathy.

Any journey, whether by air, rail, boat or bus, can be ruined by that strange creature, on whom a transistor radio is an extension of his body – either fixed to the shoulder, growing from the neck, or sprouting out of one hand or another. These are not only the younger generation either. It is a breed that has spread like rabbits in Australia before myxamatosis thinned their ranks. One can avoid such species in the open air, even at the seaside, but since they are migratory, it may be one's unfortunate experience to come across a member on any form of public transport.

They appear completely impervious to the fact that their particular brand of 'poison' may not be beloved by everyone else in the area, or perhaps the instruction book omitted any detail on volume control. I was once imprisoned in a railway carriage with a character who produced a tape recorder, and was presumably prepared to assail us all with a whole tape full of dreary pop. However, he did get off at the next station, while I was still deciding whether to make a friendly request for a bit of peace and quiet, or to tell him to turn the damn thing off!

A big advantage of travelling any distance by rail, is that one can catch up on a bit of sleep. I am able to sleep very well in sleeping compartments, and my method is to have four bottles of barley wine at the station bar before departure. This sends me straight into the arms of Morpheus for the first couple of hundred miles or so. I suppose it all depends on the measure of tiredness, whether one can sleep on trains or not. On one journey I woke up to a frosty, moonlit moorland scene, and noticed from my watch that there was still five hours to getting up time, so I turned over again with relief. Five hours later I woke again, but to my dismay found I was looking at the same bit of country as before. The engine had broken down.

It is strange the effects that train journeys, particularly at night can have. I was travelling back home from London one evening, and being a bit shattered I dropped off for about an hour. When I awoke, I took a short walk, and when I returned to the compartment it was empty. My bag and coat had disappeared, and also the passenger who had started the journey in the same compartment. Thinking he had alighted at the last station with my belongings, I rushed down the corridor, and made a proper fuss about the incident, with the Guard. On returning to my seat, there were my belongings on the rack. I had previously been in the wrong compartment altogether!

In contrast to rail journeys, which I face with a feeling approaching contentment, the prospect of flying to some destination still gives me a sense of excitement and enjoyment. I suppose it's the same for everyone, although air travel has become quite commonplace. Millions of Britons fly overseas for holidays now, and the special terms offered by travel agents and air companies put the cost within the reach of any average wage earner.

Only recently I heard that one air company are offering return flights to America for about £50 at certain periods of the year, and old aged pensioners are being offered off season holidays in Portugal, for £32 per month, and this includes their return air fare. Now this is something I heartily endorse. The more our old aged pensioners can get out of life the happier I feel about it. They are not treated particularly well, in my opinion, despite our affluent society, and the welfare state.

However, let's get back to airports.

One similarity they have with railways, is the emotionless voice of the announcer, that, unless you happen to be in the right position, seems quite indistinguishable. I must admit that I get very irritated and annoyed when she repeats time after time that the plane for which I have a ticket has been delayed, despite her polite apologies which appear to be read from a standard apology sheet. The length of delay is so vague and alters so many times that sometimes one wonders what on earth has happened to the plane.

Some years ago, Marie, myself and the family went to Manchester Airport, from where we were to fly to the Isle of Man, which is one of my favourite holiday places, and not only because of the high quality of its seed potatoes. After three hours of information that the plane had been delayed due to fog – admittedly, no one's fault – I eventually found out that plane had not even left Manchester for its previously scheduled trip, and had to fly to the Isle of Man, unload, load up again and return to Manchester, before we could board. We waited for five hours before being informed that we should all go back to the farm, and return to Manchester the next day, when they *hoped* we would be able to commence our journey.

There's no need to describe to the family man or woman, for that matter, the indescribable mental torture (and physical too), of keeping five children, aged one to seven years hanging about, full of excitement then impatience, for five hours at Manchester Airport. The new Concourse had just been opened, but the kids just about wrecked it and us too !

Airports, though, as a general rule, are far more interesting places at which to wait, than railway stations. There are aeroplanes for the children to watch, and Air Hostesses for the dads.

I have always found the observation of my fellow human beings rather fascinating, especially their behaviour patterns, which, as I have said before, are not unlike those of my cows.

However, I am not alone in this. At Glasgow Airport I once noticed two ladies, sitting opposite me, who were nudging each other and whispering, and looking bashfully across when my head was turned. After quite a while it became obvious that I was the subject of their interest, and it became quite embarrassing. I checked that I had 'rearranged my dress before leaving the room', and that was O.K., so I returned to my crossword. Eventually one of them came over and said "Excuse me, is it?". "Is it what", I asked. "Is it Harry Corbett?" she replied. "No," said I, "My name is Moult". "Oh," said she, turned her back and departed.

The system of handling baggage is one about which I have never ceased to marvel. It is a remarkable process where the bag goes on a set of scales at the departure point, and the next appearance it makes is like a dead pig in a bacon factory, travelling along a conveyor belt at the airport of arrival. I suppose that some luggage must get lost, considering the millions of bags and suitcases handled a year, yet it seems that everyone gets the right suitcase.

One would think that it would be easy to switch a bag containing valuable contraband or duty free fags, but no doubt the Customs Service are aware of every trick in the book – professional or amateur editions. I would be a dead loss as a smuggler, and would never, ever be able to pass a Customs Officer without giving the game away.

I suppose I could bring some chalk and mark my own bag, as that appears to be the only method the Customs Officers adopt to show a bag is 'clean'. I doubt even this would work, as on one occasion, just after my bag had been marked, the thing bust open and my possessions scattered all over the floor of the hall. It would have been even more embarrassing had my luggage been packed with diamonds, porn or drugs – No, I'd never make a smuggler!

There are many people who despise travel, some through the inability to be transported more than a few miles without suffering acute nausea. To these unfortunates, I can only advise they take a stronger brand of travel sickness tablet. To the others, who consider it a necessary evil to get from one place to another, I suggest they could enjoy the journey if they accepted it. It's only an attitude of mind like a lot of other things.

20

"... But Once a Year"

AT CERTAIN times of the year I get the impression that life is a huge roundabout with me standing on the edge and some demoniac being in the centre cranking up the acceleration.

On my personal roundabout there is a variety of objects reminding me of the time and date, instead of gaily painted horses, and naturally many of these objects signify some point in the agricultural year. Many more than would appear on another man's roundabout.

When it comes to December, however, and the days rush along to Christmas time, I suppose my symbols become pretty standard. There are few people whom Christmas does not affect in one way or another. In my own case there are aspects which make me feel elated, some which give me a feeling of depression, and others which I find practically nauseating, but by and large it is the happy things which are to the forefront. Perhaps there are some Scrooges around who are so cold blooded that no joyful emotions are aroused whatsoever, but fortunately I've never met any.

Christmas is essentially a quiet time for farmers, and it is particularly pleasant to have a couple of days off in the depths of winter.

The corn is in store, the potatoes in the clamps, and sugar beet delivered to the factory, which usually closes for Christmas anyway, and the landworker finds he is working more or less on his own, and to some extent missing the company of his mates.

Harvest time is past, and although through economic necessity a lot of work is performed by machinery, there is still an essential element of team work to get in the hay or cereals. During the winter the farm worker may be digging a ditch, cutting or laying a hedge (although more likely to be grubbing one out), and the stockman, who helps in the fields during cropping, will be looking after the animals. What a pity though, I can't get a few strawberries to produce about this time !

Nowadays, there are not even the plough horses to speak to. Although a tractor is a far more productive aid, it's a lonely place to be on a cold December morning.

All self respecting farmers, especially with heavy ground, like the yellow stubble turned into brown earth by Christmas. Not merely to make progress, but to take advantage of the frost, which is one of the best natural cultivating agencies.

I like a hard, frosty Christmas. A lot of sentimental slush has been written about white Christmases carpeted with snow, but that's not for me. Snow brings no joy whatsoever to farmers, animals or for that matter anyone else, except a would be skier on some Pennine slope, providing, of course, he can see his way to the bottom through the accompanying fog.

The countryside has a distinctive beauty in winter, increased by the many sounds which can be heard in the clear air, sounds which fail to carry in a drowsy summer atmosphere.

One's senses seem particularly keen when standing on a dunghill on a frosty December morning. It is a sensation that no one could appreciate unless they have experienced it. At night, against a background of nature's sounds, one can hear the continuous hoot of the brown owl mingling with the bark of the dog fox as he seeks his mate. There is an old legend which tells that animals, certainly cows and donkeys have the power of speech for an hour on Christmas Eve, but this is something I have yet to experience!

At one time when the pleasures of man were more limited in number, eating and drinking were relatively more important than they are today, and therefore the Christmas Fairs and Fatstock Shows became firmly established.

Of course, in modern times most of the produce flows away from the farms bypassing these ancient institutions. Yet they still have an attraction, judging by the way the butchers bid for the Champion bullock, pig or sheep. Presumably the customers will all be dining from the winner of the rosettes proudly displayed in the shop. Not all the animals on show are slaughtered, and dark secrets are whispered about how certain competitors brought an animal from

Birmingham to win at our particular show.

One change on the farm I have particularly noticed, is the absence of poultry. I dropped poultry some years ago, again for economic reasons, but some of my early memories of farm life concern geese, turkeys and birds of every size and description – edible of course !

During the Christmas period we wallowed in a welter of feathers, gore and guts – anyone who could and would pluck poultry, plucked with a will, and some of the really skilled performers could strip a cockerel in three minutes, without damaging the flesh. This was only possible though, if the temperature of the bird was 100°F, which was only slightly less than normal, so we sat on a crate, pulled out a bird, killed it quickly and had its feathers off before his wings had stopped flapping.

Geese and ducks took more trouble because of their down. The feathers were a saleable commodity, because not all eiderdowns are stuffed with feathers from the Eider duck.

I, for one, welcomed the disappearance of this performance. Our birds are now plucked, dressed and practically served up, in a clinical, highly automated plant. Many of course are frozen, but does this matter so much? I doubt whether anyone can taste the difference between a frozen turkey and a fresh turkey. I certainly cannot.

One result of the new methods in poultry production is the avoidance of the rush of former days, wondering whether to execute the birds while the weather was mild, or whether it was too early, or to leave them, chancing that no really cold weather was on the way. Now the turkeys on the majority of Christmas tables will have ended their days while their purchasers were still enjoying summer holidays.

Fewer people now work on Christmas Day. Postmen have been excused, publicans open if they want to, so there must only be nurses, clergymen, milkmen, farmers and to some extent railwaymen who do any. Oh! I nearly forgot housewives.

If a farmer keeps cows, or any other animals for that matter, these must still be cared for during any holiday, and cows need milking twice a day whether it is Christmas Day, Bank Holiday Monday or Shrove Tuesday.

However, the stockman's job is far less arduous than it was. Unless an uncooperative cow decides to have a calf, today's methods make his life much easier. When cows needed to be hand fed, the stockman would spend a great deal of effort during the days approaching Christmas, mixing huge piles of food, and gathering piles of fodder in strategic places, almost as if he were gathering ammunition to repel an attack. This 'ammunition' would then be discharged into the cows during the two days. He was also somewhat of a superior lavatory attendant to his charges, and it was a common sight to see him desperately pushing his barrow, brimming with muck, up a steep plank to the top of the muckheap, trying to get home to his Christmas nosh on time.

Self service feeding arrangements now relieve him of this duty, and the yard only needs cleaning once or twice a year, and this is done with a tractor and loader.

There's one thing we can be sure of, and that is that there's no danger of us getting the date mixed up, after several renditions of "Good King Wenceslas" and "Silent Night" on the milking parlour radio, followed by the constant reminders on the telly. Perhaps our sharing the entertainments provided on these media have, to some extent, replaced the party games and fun that used to be the essential part of Christmas night.

There are people who insist that Christmas is not what it used to be and never will be, or that it has reverted to a pagan festival and that we forget whose birthday it really is.

It *was* a pagan festival for thousands of years before Christ. The Winter solstice, so soon to be followed by the lengthening days and the faint promise of Spring, was a natural time for rejoicing.

In the North especially, where the winter is hardest, it expressed men's longing for the return of light and warmth, before central heating, and also for the return of vegetation to the fields, likewise before quick frozen vegetables. The return of heat and light must have been eagerly awaited if one spent December in a cave munching acorns.

The Roman Saturnalia began on December 17th and continued for seven days when those who presumably had not retired hurt, were exhausted. Gifts were exchanged at this period, and the whole idea of new beginnings, plus the vision of the long days just round the corner was so deeply engrained in the tradition, that when the Church sanctified it, and replaced the Sun God by the Holy Child, it ensured its permanence.

I suggest that it is really for the climatic reasons that in this country, most people would think of Christmas as the greatest Christian festival, and not Easter as in the case of warmer countries, which surely has the first claim in the theological sense.

One thing I do dislike about Christmas is the preoccupation, more accurately the obsession with food and drink. You may think it strange that I should say that, being a farmer, for surely anything that increases the consumption of British food is O.K. by me. I get a bit tired however, of the telly commercials, not for advertising the stuff, but for the way that they seem to presume that nobody gets a square meal at any other time of the year! There is no doubt that we do stuff ourselves too full – the indigestion pill firms must have *their* harvest around Boxing Day! Perhaps people live on bread and marg. in January. Certainly the quiet markets at this time give some support to this theory.

On the other hand I also get a bit tired of being told that Christmas is too commercialised, whatever that means. Do the pundits mean that life all the year round is too commercialised, and that Christmas time should be some kind of isolated euphoria when the husband and his spouse kill the fatted calf of their own feeding, and ignore the fruits of modern progress like frozen veg. for the duration of the period?

Christmas cards also irritate me a bit. I used to think that Christmas cards were a poor substitute for a letter, especially to an old friend, whom one had not seen for a long time. Now the whole thing has got out of hand.

One might say that the business of sending cards is the inarticulate, relaying prefabricated messages at high cost, but that would be a little hard; nevertheless it does seem a grinding chore, and so expensive too. Still, it is encouraging to see more and more charities deriving benefit from the sale of cards. How convenient too, and how democratic, to send an Oxfam or some suchlike card, instead of wondering if the one that you can afford will look big enough on your neighbour's wall!

Christmas cards are, I suppose, like work; rotten to be without yet capable of being overdone. How sad for instance, is the sight of an almost empty mantlepiece in a pensioner's cottage. Are old people neglected at Christmas time? Well. I've been told that the intake of aged relations into Old Folks' Homes reaches its peak at about this time. Surely this can't be true!

Since we all live in the age of mass media, every experience can be readily shared, but a lot of the old customs have disappeared. The mummers, for instance. I have heard the old people talk of the time when they came round to do their act. I believe that the charactors varied with the location, but the theme of death and resurrection never changed. There is a fight between the main character, usually St. George, or King George and an adversary; King George is killed; a doctor revives him and he is restored. Perhaps this is representative of the seasons, the rise and fall of the leaf or even some primitive vegetation rite.

I hope the reason behind Christmas isn't disappearing. There's another birthday I never forget at this time of year and that's my wife's – on Boxing Day!

21

Farming the Moult Way

It must be a sign of old age when one realises how time passes, and perhaps it is a sign of hardening arteries when one finds a resistance within oneself to change.

Only the other day I was looking at my milking parlour, and realised with shock that it had been installed for some fifteen years. What was one of the most up-to-date, abreast of the times systems has now been superceded by the tandem system, the herringbone system, and now the rotary system, in which the cows circulate around the operator like a huge dart board, needing a minimum of manpower per cow.

When I came to this farm I decided to keep cows in yards. At the time the common practice was to tie them up by the neck in cowsheds, for the duration of the winter, but this meant that the farmers had to go round the sheds to check whether any of the cows were hanging from their ropes or whether everything was as it should be. The local vet used to say that if no one rang him up ten minutes after the television programmes finished, which in those days was about 11.00 o'clock, he knew he'd have an undisturbed night.

One of the joys of the loose housing system, is that this chore of keeping an eye on the cows, although desirable, is less necessary than under the old system. Unless a cow is actually calving, it is unnecessary to look round every night. Even if one is calving, at least it's not in the same predicament as one tied up in the cowshed, unable to reach the newly born calf if it happened to produce one without proper warning.

When asked do I keep cows, I should reply, "No, cows keep me", and this has been consistently true, up to now anyway. Of all animals to benefit from walking about in grassy fields, I'm sure cows are the most rewarding, although they do need milking twice a day, including August Bank Holiday, Christmas Day and Boxing Day.

Cows eat two kinds of food – concentrates, or roughages. Concentrates are cereals and nuts bought from the corn merchants, which are very expensive and getting even dearer. Roughages are grown on the farm in the form of hay, silage or even straw, or alternatively bought, such as brewers grains – like the brewer's tea leaves after the barley has been made into malt, which is a very useful food for cows. It may also be a root crop like potatoes or sugar beet pulp, and it seems sensible to me, that if you can give the cows high quality roughages, you can expect a better job of making milk as regards profit, than giving them low quality roughages and trying to compensate with very heavy feeding of high cost concentrates. So I like a cow with four good feet to walk about on, and a big belly to eat and transform the roughages, which I hope to make, into milk.

I have always been a silage addict. The best description I can think of for silage is 'pickled' grass, unlike hay which is dried grass. My farm sometimes smells like a pickle factory when the silage is fermenting, and can get embarrassing to the people who work in it. You can't smell silage yourself after working in it all day, but there's no doubt that other people can. A friend of mine called George, used to travel home on a bus after feeding silage to cattle, and wondered why all the other passengers rushed to the other end when he got on.

I was forced into silage, not through economics, but by hay fever. I had miseries trying to drag myself round the hay fields in the old days, until I discovered that with a forage harvester that gobbled up the grass, just like a Chancellor of the Exchequer gathering taxes, and blew it into a trailer, most of the dust had disappeared before it got into the yard.

Another advantage of silage is that it can be arranged so that the cows help themselves, as in one of those new rail bars on Paddington Station. This saves enormous time in the winter, although it is important to make sure that the silage bar stays open long enough for the more timerous ones to get their fair share.

Cows have a habit of producing calves, which is a prerequisite of giving milk in any case, and a Friesian cow crossed with a Hereford bull, makes a jolly good by-product for eventual arrival at the steak houses. The easiest way to rear calves is to let them suckle the cow, but if, like me, the farmer wants some milk from the cow also, the calf is often taught to drink out of a bucket, making sure, of course, that it has its mother's milk right from the start. This is usually my personal job, but very wearing on the teeth. While I'm wrestling with the calf, trying to stick its head in a bucket, and avoid drowning it at the same time, my teeth have a stupid habit of holding a pipe. When I grit my teeth with the effort, the pipe stem usually breaks, but on two occasions it was the teeth that went.

Come to think of it, if one must smoke, a pipe must be the best thing on a farm. I did try cigarettes once, but the wretched smoke got in my eyes, whilst picking potatoes. At harvest time, one can fix a pipe up with a mini-dustbin lid, which enables one to puff away merrily, without endangering the whole field with fire.

From the age of 17 I've been a pretty consistent pipe smoker, the highest point of my smoking career being when, for two years, I was in the last three of the 'Pipe Man of the Year' competition. I have on the mantlepiece at home the two mini-silver pipes, with which I was presented. Pipes do have their disadvantages. Wear and tear on my incisors, coupled with the strain of biting on the pipe stem while trying to dose a cow with a bottle of medicine, or feeding a calf, resulted in my dentist supplying some special re-inforced gold teeth, which are, as my wife says, a credit to the National Health Service.

I live on a mixed farm, on which cereals are grown too, so the cows lie on a bed of straw in the winter time. Of course, this is not one of the most modern systems, compared to the cow cubicle system, where a cow stands in a little stall with its back end hanging over a central alleyway. The theory is that the bed keeps clean, and the gangway receives all excreta which is pushed into a heap by a huge squeegee, usually mounted on a tractor.

With my old fashioned system, we don't have a problem with all this material, which is called slurry. Very often, in a modern cow unit, the slurry takes more dealing with than the milk, and so far, this material is an embarrassment. If anyone knows a good use for it, I'm sure there's a good prize waiting for them. In this highly scientific age, maybe we'll be growing some bugs on it, or reprocessing it before very long.

One of the dangers of cow keeping is the disease called bloat. This occurs usually in the Spring, when a cow filling its inside with some gaseous material like clover, finds that the gases inside the rumen expand to such a point that the diaphragm which is necessary for breathing, ceases to work, the cow becomes distended and blows up, and actually dies through lack of oxygen because she can't breathe. Our quick remedy is to use somebody's black drink, or red drink, or peanut oil, or even any kind of detergent. A packet of Daz is fine, but in dire cases it is necessary to puncture the tightly stretched skin between the hip bone and the back rib and let the gases fizz out. Now the correct way to do this is to use a trocar and canula, but as this is vet's apparatus, on one occasion I had to improvise. You've heard that saying "To save the cow's life, he had to cut its throat", well, it was similar on this occasion. I stuck an old bayonet in the cow and out came the gas like a kettle on the boil. The cow recovered, soon healed up, and everything ended happily ever after.

Like human beings, cows vary a great deal in their nature. We had one called Blodwyn, which, as the name implies, came from Wales. Often, in a bunch of heifers, there is one, called in Derbyshire "mardy". That is to say, very quiet and tame. I expect that in coming from Wales, this particular bunch had been reared with loving care by some farmer's wife or daughter, and consequently were so used to humans. Blodwyn would come round and look over the garden fence in a very pet like manner, and she used to give the kids a ride on her back. I cried when Blodwyn had to go, but economics have to overcome sentimentality in farming, and farmers are notoriously hard headed.

Another crisis in cow keeping comes at calving time. Most calves are born naturally, and the main thing is to ensure that the calf is in the right position before it is actually delivered. If it is head first and its feet are in line there's not a lot to worry about unless it is a very huge calf and a small cow. However, when a calf is presented hind legs first as it is born, the umbilical cord is severed and oxygen is denied to the newcomer, so it is very important to pull – quickly. There is only about 30 seconds in which to get the calf's head into the fresh air, otherwise it will die from suffocation. I've seen many a calf thought dead when born, and I think it interesting the method used to get it to breathe again – put a sharp piece of straw up its nose to make it sneeze, but best of all hang the calf over a door so all the mucus in its lungs drain out.

It is, to me, a moment of wonderment when a calf, teetering on the verge of life and death, suddenly makes up its mind to live. New life is a wonderful thing. As my cowman says, "In moments like this, it makes me come over all religious".

Milk fever, another curse of the dairy farmer, is due to a temporary lack of calcium, and causes paralysis. It can easily be cured though, if discovered in time. Sometimes we separate into another field the cows ready for calving and which are usually dry, that is, not being milked. They need careful supervision and inspection because not only could they have something wrong with their udders, such as mastitis, which is quite common, but they need checking about twice a day to see if they are healthy in other respects. We had a real shock about three years ago. It was a memorable day really, because it was also the day for our cows' bi-annual test for T.B. to which we were all subjected. All cows in the country are now attested. We rounded up all the milkers, but when we went to round up the dry cows, which were on their own, to our horror we found one was dead, yet only the day before it was all right.

The first thing I worried about was anthrax. It's no use poking around a dead cow in case it has this rare, but deadly disease to humans, therefore we looked at it from about a hundred yards away. This was a silly thing to do really, because it transpired that if a closer inspection had been made, we would have discovered the cause of death, and saved ourselves a lot of worry. The knacker man, who had taken the cow away, told us two days later, that there was no calf in the particular cow. No one had noticed that she *had* calved, so presumably milk fever put paid to her. Then I thought "Where is the calf?" I hate mysteries, so immediately went down to the field and began to search, but to no avail, until I saw a black, woolly, curled-up object near the ditch. It was the missing calf all right, which, even after two days in the field, was so vigorous, it put its head down and ran all the way home. This was some consolation for losing the cow.

I quite like milking, and this is the time I get some of my best thoughts. Some thoughts are second hand, of course, but there's no originality that's not undetected plagiarism. One of the compensations or delights of milking is to have the radio on in the parlour.

Some people think that cows like music, and I tend to agree with them, yet I don't really believe that cows can tell music from rubbish. The principle is that if you maintain a certain level of noise, any sudden clatters or bangs are masked, and therefore cows or pigs or hens are less nervous in disposition. I think this applies equally as well to all types of animals.

When it's my turn for morning milking, I listen to the fatstock, thinstock or any other stock prices, cereal market, that smashing chap from the Manchester Fruit Market, Bill Nicholson, with his funnies about prices of cauliflowers, Worcester Permains and Grannie Smiths, and then, of course, the music. Sunday mornings are particularly good. If I've missed Alistair Cooke on Saturday night, I can hear him on World Service, and the cows really appreciate the news, whether English or Polish, and what used to be called *Lift Up Your Hearts* or its modern day equivalent. Then cows are funny things. Only the other day, as I was fetching them out of the field to milk, I saw one obviously with some obstacle stuck in its mouth, and it was coughing and slobbering, and grinding away. Eventually I found it was chewing an old bone that the dog had left. Getting its own back, so to speak!

Bulls are supposed to be fearsome creatures, but we had a Hereford bull which needed protecting from the cows, as every time we let him into the field, the cows knocked him about. At the time I had a big wooden hay trough, into which the cows bundled him, and I had to saw the thing to bits, before I could get the bull out.

Cows do daft things too, like getting upside down in a ditch, which is a pretty sight to see when you're fetching them in. They begin to swell up after a few hours if no one notices, and get wedged in the ditch tighter than ever.

I've fed my herds on parsnips and carrots and brewers grains, and as regards a musical note, my mother gave me an old Post Horn years ago, and this has been invaluable. A blast on this and the cows belt up to the gate in no time. They can be scared too. I've seen them bellow with fright at the sight of a mouse in the trough, or an odd shaped plastic bag, before they devour it in an attempt to commit suicide.

Whilst I was testing the cows one day, we had a visitor – a surveyor chap – and this was the bombshell he dropped: "Is this the farm the motorway's going through?" he asked. When I picked myself up off the floor I said "What on earth are you talking about?", this being the first intimation I had of the new M.42 Motorway, connecting Nottingham and Birmingham.

This, of course, was a terrible shock, and its proposed construction has been hanging over our heads like the Sword of Damocles, ever since.

We live on the edge of a valley with a breathtaking and unobstructed view of the Derbyshire countryside. The thought of great pantechnicons, lorries, coaches and the like belting up and down across the cow pasture, seems like desecration.

I suppose living in the Midlands, one must be close to the hub of things, but one didn't bargain for these radii which are part of the Motorway network. In any case, South Derbyshire has had a pretty fair bashing through 'development'. In the first place we had all the major coal fired power stations along the Trent Valley, which was convenient to the coal mines, with all the attendant overhead cables. When they were first put in, they probably represented power and progress to the countryman, and it wasn't until they got well under way that we started yelling. It is usually townspeople who shout the loudest about the preservation of the countryside, because the countryman often doesn't appreciate the excellence of his own environment.

So as I write these lines, the proposals are to cut this motorway through the farm, to ensure that people living in Newcastle-On-Tyne can rush down to Bristol with the most expediency. It's most ironical, and even seems hypocritical to see a man speeding down the M1 with a sticker in his car saying "stop the M42".

I certainly think the shortage of fuel, the criminal neglect of the railways which should have been made to carry more goods and therefore justify themselves, and piecemeal planning of motorways, should all be investigated in a far more comprehensive manner than is being carried out at present.

What can the farmer do about it? He can protest, he can band together with his fellows, he can lobby his M.P., and all those things have been done. In fact, I went to the House of Commons to lobby my own M.P. He was very kind and treated us to a tea of, guess what – strawberries and cream – on the Embankment. Later, I was amused to read in the local newspaper, that Ted Moult and a crowd of villagers from Ticknall had protested against the new motorway, as if we had rushed back up the M1 to dance around the maypole like some creatures out of *Schwander the*

Bagpiper.

The awful thing about motorway protests is that if one says "I don't want it", it is given to someone else, unless it is stopped altogether. A lot of the protests are selfishly motivated, which is reasonable, I suppose, but there is, I would add, a responsible body of opinion, who would query the whole development of this island before it is covered in a sea of concrete.

Surely, Britain has enough of this at present. Are we to speak out before it is too late, or surrender more of our land to the motor vehicle, for ever? Mass transport by rail is still possible – that system just having escaped complete dismantling. Public resources should be put to that end!

It's time the complacent public became alerted to the insidious pollution to the ecology, that a short sighted commitment to further motorway construction would mean.

Not all our visitors are bearers of ill tidings, in fact, at Scaddows Farm we have, from time to time, entertained many groups of people, different in age, jobs, and even race. Entertained may or may not be the right word, but I suppose a bit of extra knowledge has been imparted in one way or another.

There appears to be less ignorance than hitherto, between city dwellers and the farming community. Whether this is because of *The Archers* on radio, or other programmes on the telly I don't know, but perhaps in a humble way, farmers have managed to put their side of the picture more clearly. Certainly it is not for the want of trying, as the efforts of the N.F.U. Information Department will verify.

Thinking back, surely all those meetings with Women's Institutes and Townswomen's Guilds had some effect. I prefer women's meetings to the male counterpart, as they appear so much more polite and civilised.

Probably our most appreciative visitors are junior school children, slightly more advanced than the *Playschool* vintage, who just love to fondle the smaller animals, particularly the lambs, calves, and even kittens. They gaze in wonderment at the big, flashing combine harvesters and tractors, thoughtfully painted in bright colours by the manufacturers.

I know a farm where visitors can watch the cows being milked, and I reckon this is not a bad idea, yet the thought of milking with a crowd of human goldfish squashing their noses up again a sheet of plate glass strikes me as hilarious – for the cows ! !

Each year we seem to have a Farm Walk of some type, and it seems that most of the local Young Farmers' Clubs have seen what we are doing, and what we are trying to do. It's an odd fact, yet rather a pleasant one, that farmers at all levels, and as far as I can tell in all parts of the world, are quite prepared to show each other their individual methods.

No doubt some of this free dispensation is good for their ego. I doubt that farmers realise they are in competition with each other, which they are, of course, but not to the same extent as a greengrocer or a hairdresser. I wonder if the captains of industry adopt the same attitude, or is this called industrial espionage?

Personally, I like the idea of young farmers, or young anybodies, for that matter, visiting various farms. Not that all are intent on improving their knowledge.

The first time I was asked if I could conduct an organised tour round my own farm, I went to considerable trouble to lay on food and refreshments, and offered to give a short talk at the end of the tramp round. Nowadays we don't go to that much trouble, just direct them to the local boozer if requested – it's less bother.

However, on that first occasion, as we all sat in the swept out tractor shed after I had finished droning on about fertilizers, silage and the like, I foolishly asked "Any questions from anyone?" "Ah!" said one superyob "Any more coffee left?"

Non-typical of my usual 'clients', that boy must have been the rudest; the most appreciative was a party of handicapped children, and the most exotic was a party of ratings from the Kenyan Navy.

In addition, there have been visitors from the Institute of Bankers (why shouldn't they see where their money went?), and assorted Estate Agents and Young Wives, though not at the same time.

One type of visitor who has invaded my home, farmyard, time and even thoughts since I have been farming, is the commercial representative. Yet it's somewhat incorrect to bracket them all under a type. They have distinctly individual approaches, and personalities, but one thing they have in common is an irritating tendency to 'just drop in'.

Travellers can be placed in three categories – the locals calling once every month; those that call once every year, and those working on the Phineas P. Barnum philosophy that there's one born every minute.

Their deplorable practice of 'dropping in', can only be excused in those calling annually, in my opinion. These are principally debt collectors anyway, and I have no doubt that the techniques employed by certain farmers to avoid the payment of overdue accounts can match, in ingenuity, those used by the collectors.

If I were a traveller visiting a farm, I would certainly do it by appointment or arrangement, or at least drop him a postcard telling of my estimated time of arrival. A casual call could find the farmer at work which is probably absorbing his time and interest, he could be at lunch or in bed, or even in a situation of embarrassment, should an unexpected caller arrive.

It is an absolute waste of time and effort on a traveller's part to coax a sale out of a farmer who has other things on his mind.

A very earnest young man once wasted his own talents in telling me what an idiot I was not to grow his particular firm's seed, while I was struggling to get the milking machine on to a fractious, newly calved heifer.

Most farmers do appreciate the part commercial concerns play in the scientific development of agricultural ways and means, and are usually only too pleased to have a chat with a fellow who knows his subject – but, at a mutually convenient time.

Not all farm salesmen follow the scientific path, however, in fact I have known those who were more akin to witch doctors. These peddled those ubiquitous drenches they called animal medicines, guaranteed to cure all known, and some unknown diseases. On one occasion, a lorry immaculately painted in canal barge style turned up at the farm. The driver made some most extravagant claims for his load of crystals, which transpired to be common salt.

Most companies who depend on human representation for their products must realise that the factor which initially impresses a prospective customer is their salesmen. However good a product, or however cheap, it is he who gets his foot in the door, and more times than not clinches a deal.

I'll not deny that sincerity in such men is essential, but I get the impression that some are brainwashed to such an extent that they actually believe what they are telling you. Mind you, this is a far better approach than one individual whose opening gambit was "This stuff isn't much good, but I have the six children to feed".

Then there is the intellectual type, who readily drops out quasi-scientific terms, in an attempt to baffle the customer. I suspect the salesman is baffled too. Terms are used freely, which are so indefinite that they become meaningless, or in some cases ludicrous. "Pedigree, field inspected, once grown", whatever that means, are all phrases which have been flung at me over the years. I have even taken a leaf out of their book by once selling some surplus potatoes as 'guaranteed non-certified'.

I remember a salesman trying to sell me some organic fertilizer. He claimed that the organic content was the finest point about it, but I argued that ten tons of his product could contain only a fraction of the organic content of a nearby load of muck. Moreover, it was a sight more expensive.

"That may be true", he said. "But even if our stuff costs more, the best is always the dearest (a standard cliché), and it is available over a longer period". "Would it not be better, therefore, to make several applications of a more economical fertilizer", I retaliated.

"Well," said the man, "Mr. Giles in the next village buys it, and he's no mug." I found his argument inconclusive, and resented the inference that I was an idiot, however well meant. Needless to say he didn't sell me anything.

One salesman, who can visit my farm any time he likes, had a most impressive eagerness for hard labour. He would breeze in at the double, wrest whatever tool I was using from my hand, and do his best to lighten my load. Once he rolled about six acres before he departed with a cheery "Nothing you're wanting from us, I suppose?" I liked him, and often wondered where he went after withdrawing from the selling business.

Perhaps I have been a little harsh in my views, because I know that for some, the salesman is a guide, philosopher and friend, and sometimes a shoulder to cry on. On the other hand he is often an Aunt Sally for those who wish to poke fun.

When the city dweller thinks of farming, if he ever does, his first thoughts are probably of hay making or harvesting.

In the old days we made hay in the field, swept it up to a stationary baler, then stacked the bales and carted them off. Now, of course, over the last 20 years or so, pick-up balers, which require no explanation, do the work for us. But in those days, the hardest work was to pitch the hay with a long fork, into the stationary baler.

I well remember one day when the hay was ready, the baler was ramming the bales together and the sweeper under the tractor was pushing great loads of hay towards us. It was very hot, and being a considerate employer, (not that I was employing many, but I had a lot of friends who came to help), I asked a visiting friend to go up to the farmhouse, where I remembered I had bought a cask of best Hereford Cyder – that's what the advert said. Instead of the silly chump coming back with a jug or so, he brought the barrel, and set it up on one of the empty bales.

Whether it was the hot day or the lethal qualities of the scrumpy, I'm not sure, but after two pints of this heady brew, I didn't know whether I was throwing myself, the hay or the fork into the baler, and old Dick, who had come along with his horse and hay rake, somehow got his horse upside down in a ditch on the way home. Next time, I thought we'd stick to a more innocuous kind of beverage.

Corn growing used to be a straight forward thing. Even in my career I have sown it out of a hopper, just as they did in the Bible. We used to mow it with a binder, tie it up in sheaves, stook in the fields, and carry it. The best job was pitching it onto the wagon, the lad's job was making the loads, and the real skilful job was building the stack. Then in the winter we had thrashing.

I used to grow sugar beet as well, but I couldn't grow both sugar beet and potatoes, and the latter got my vote. With potatoes I can see the customers.

My farm's a late farm, so it's useless to grow early spuds. I grow main crop. Being late I miss the best price, but that's how it has to be. I must be doing better on potatoes, because they are harvested into big boxes holding about five hundredweight. I pay the pickers 25p a box, and as this hasn't gone up since 1966, I must be growing more to the acre, which makes picking easier, so the pickers earn more money. I like spud picking, for not only do I meet interesting people – some more interesting than others, of course – but if you can get pickers, it's better than having a harvester, because all the harvester does is to carry the pickers around. It's also difficult to pull harvesters up and down hills on a wet season.

About ten years ago I had a notion of buying the farm next door, and even went through all the staff work with the bank manager. He said it was a good idea, but it was too much for me. In other words, "you can't have the lolly, Ted". I decided, by way of consolation to grow a couple of acres of strawberries. I discussed the matter with my friend Graham Ross, who was in the Advisory Service at the time, and Arthur Pearn, who still is in the Advisory Service at Harrogate, as far as I know. They said that thinking about a job is not half as difficult as doing it. What I had to do, was to buy some plants from the Ministry of Agriculture book, listing those with a Health Certificate. It is very important to get plants with a Health Certificate, because virus disease in strawberries can make the difference between chalk and cheese, to use a cliché.

I asked "What do I do now, Arthur?" He said "Clean that field up, plaster it with cow muck, and you need plenty of potash, then plant your plants." So we bought the plants from Norfolk, planted them a yard between the rows, and half a yard between the plants, and they grew well until they had a nice lot of flowers.

Arthur came along later, and I asked him what I did next. To my dismay he said, "Pull the flowers off". He explained that the first crop is lost in the hope of getting higher subsequent crops. So I learnt that strawberry growing is long term. One has to be particular to get a clean field, pick the flowers off the first year, and hope you get three reasonable crops before the constant battle with the weeds is lost.

The great attraction about strawberry growing is that there's no great capital outlay, although speculative. In a wet period one's potential profit can evaporate like margarine in the sun. It can use one's capital equipment, more or less, but it is a job that requires timeliness more than most other crops grown. Tomorrow won't do, and one has to have a special programme of sprays etc. When I say sprays, don't rush away in terror against the word, because a lot of them are used, not to kill insects, but to protect against grey mould and mildew, which are fungus diseases.

One is told by old people whose taste buds have possibly become atrophied, that Royal Sovereign is the bees' knees in strawberries, but personally I find it a slightly acid taste, and Cambridge Favourite seems to be *the* favourite, in my experience.

For the first few years we grew strawberries, and it provided work in July when I should have been on holiday, we relied on pickers, due to the foresight of the Leicestershire Education Committee which allowed their schoolchildren to break up at the end of June. The fruit was then sold at the market on commission.

I wonder whether there was any political significance in the year of the election, in 1970, because the strawberries ripened with astonishing alacrity. It's very odd that, because, for ten years I have grown strawberries commercially, one could predict that the date of the first pick in my area would be on June 30th, and the highest total yield would be July 12th, or Bastille Day. In 1970 they were a fortnight early, and in 1972 a fortnight late.

One of the results of the early season was that we had no pickers. We had already tried a system of letting the general public in to clean up fields, after the main crop, right from the beginning, and this had become so popular, that we decided to rely upon this system to get most of our crop picked. We now provide car parks, last year we sold cream, signed autographs, and I was amazed to see the numbers and types of people who pick strawberries.

It is fascinating to see the public at work, enjoying themselves, for me. Coming into the country, the air's free, the view's free, they pick what *they* like so it's *their* choice, they eat what *they* like, we hope *they* don't tread on too many plants. As far as I'm concerned there are no bad debts, no complaints, what the customer doesn't like he doesn't pick, there's no packaging, and there's no transport. All in all, it's a most excellent way of marketing.

One day last year we had about 300 people in the field and a sudden thunderstorm made them all bolt through the hedge like a herd of charging buffaloes, and they disappeared into their cars. The rain came so quickly they didn't bother to pay for what they had picked.

It's a nice job watching the public pick strawberries. I get Charlie or Bill or George or Frank to take the money at the gate, but it's a busy time for all of us. We do get our moments of drama, though. The objects that people lose more than anything, are their ignition keys, but one poor girl did lose her engagement ring. Last year, the total tally was nineteen ignition keys, one engagement ring, three purses, and one set of false teeth, otherwise everyone came out intact.

I've only mentioned the things that I am doing currently, or have done in the past. Who knows, situations change, and may lead to alternatives. I'm considering, for instance, growing rape for oil seed.

Beef is an attractive prospect for people who like to walk around their fields with a dog and a stick, looking at some nice, contented, sleek Hereford bullocks, but for me, it's never been profitable enough. For a start you need so much capital in beef, to produce a profit. I can't see the sense in buying store cattle at 24p a hundredweight and then selling later in the summer, when, admittedly they are bigger, to make 20p a hundredweight.

I used to have sheep, and I suppose I could get very interested in them again, but sheep and cows mix only when they are kept well apart. The sheep eat all the best grass, fencing is a big problem, and there are also dogs. Anyone who has never seen a dog worrying sheep could not start to realise the horror of such an attack. I've seen a dog worrying my sheep. At the time I had a boy working for me, and sent him back to get my gun, and shot the Alsatian dead. When the distressed owner appeared, all I got was a stream of invective and later in confusion, I was accused of pulling wool across the dog's teeth – or wool over her eyes!

Hens and pigs in the main must be considered separately, because hens can be kept in an old textile mill, and pigs are extensive husbandry enterprises. I always thought I had a lot in common with pigs, but I haven't had any on the farm for the last fifteen years. I have diversified as much as I want to at the moment.

It is necessary to employ oneself and to even one's labour requirements out as much as possible throughout the year, and to utilise each piece of equipment to its best advantage.

This means ensuring enough work for, say, a big combine harvester to do. With this in mind I managed to purchase some land in Bretby Park in 1966 – three miles away. I run this holding in conjunction with my tenancy at Ticknall, concentrating the milkers and strawberries at the latter, and the cereals, potatoes and certified strawberry plants at Bretby.

Farming is as much a business as textile production, engineering or anything else.

A farm is not the retreat for those seeking retirement, it is a way of life which involves a deal of hard work, but to me, brings the ultimate in satisfaction and peace of mind.

22

How's Your Quality of Life?

THERE is an old adage that some people know the price of everything and the value of nothing; and I believe this to be not far from the truth. It seems to have some bearing on an individual's capacity to make money. Of course, the chaps who make the millions include all types – civilised, uncouth, jolly, miserable, broadminded and introverted.

However, I am sure that all have one thing in common. It's hard to define, but I think that those who *are* finanically brilliant have some kind of extra-sensitive brain cells perched somewhere inside their cranium. Looking back at some of my former school chums, I find it a revelation to discover that the lads at the lower end of the class have often done better for themselves than some of the potential brighter material. Perhaps they have that quality of singleness of purpose, and, relieved of a wider interest in some of the ingredients that make for the much advocated 'quality of life', they have been able to plough on, thinking and acting in straight lines to their pecuniary targets.

Another factor common to successful chaps, and it applies to politicians too, is strength of constitution. What spurs them on? Money? Power? I suppose so. Who said, 'Power tends to corrupt, and absolute power corrupts absolutely; Great men are almost always bad men?' I'll tell you. It was the nineteenth century historian, Lord Acton. Perhaps it is a bit cynical to believe that statement entirely. Maybe we should mitigate the cruel words, by thinking that ambitious men (and women to some extent), have a combination of desire for self-gloritfication and wanting to serve their fellow men.

Why is it that so many of the lucky brigade who are able to rise to the top of their particular ladder, spoil their natures with a fierce, aggressive attitude to the rest of us? I say lucky, because we are all captives of the effects of our heredity and of our environment which, after all, is, or was, in the lap of the gods.

A common feature of the behaviour of those who have made it, is to surround themselves with possessions as if to prove to themselves, and others, that they are really something special. Take a typical case of a man who has made a lot of money with the help of those imaginary brain cells, and yet needs a Mark 10 Jaguar or, better still, a Rolls Royce to bolster his ego.

It's perhaps significant that there is an apparent need for those who can afford it, to live and act in a different way from the common herd. They may send their kids to fee-paying schools (especially if they went to the elementary variety themselves), set up coats of arms, drink different drinks, or whatever they think can maintain their position in society, and to act with hostility to all those who they think may represent a danger to their way of life.

I am not against private schools as such, in fact at one time some of my own children used them to good effect. As I write, all my lot are educated free of charge – free, ignoring the taxes that pay for the system. It's fine to have the privilege of choosing a particular school for one's brood, but to talk about the principle of having a choice can only be academic, as most of us are governed by economic restraints.

I have always considered that the correct way to reform the educational system is to make the State schools so good that everyone had the desire to avail themselves of them. Yet I certainly wouldn't abolish private establishments, even though some seem to be living hand to mouth and having a flag day every time they buy a new microscope! Neither would I abolish anything which functioned in any context (unless it was patently evil), if I could not substitute something better in its place.

What do we require in a school anyway? Correct me if I'm wrong, but education comes from the word educe, which according to my dictionary, means to draw out, to bring to light, to evolve. Talents should never be left to lie fallow, but brought to the top. Like cleaning a field on the farm, sometimes a variety of implements needs to be used to ensure the best results according to the circumstances. Bringing out the best in a pupil must be an exacting task for the teacher and one cannot be surprised at the occasional outburst of frustration manifested in, for example, comments on school reports. The best one collected by our kids so far is: 'His work varies from downright disgusting to barely acceptable!'

Companionship is the next consideration. No one wants to send his kids to a school full of little cheats or tarts, even if they got that way by example (or lack of it), and not from being congenitally wayward. No good can come of a lack of discipline, and imposed discipline is well justified if it engenders self-discipline. In our family it takes less effort sometimes to do a job oneself than motivate the children into doing it, but that, I would say, is the easy way out – which I often take.

Having been to a day and a boarding establishment in turn, I consider the latter has definite points in its favour. Unfortunately, in our society it is often the wrong children who benefit. If the schools are *in loco parentis*, orphans and children of criminals, lunatics and absentee parents must get the most out of being boarders.

I believe too, in the principle of comprehensive education, if only because it abolishes, at least in theory, the application of labels at the age of eleven. An eleven plus failure is virtually barred from so many avenues of progress. This must be bad. Criticism of the size of these schools is often voiced, but the main objection comes from those who see the creation of neighbourhood schools, each reflecting the prosperity of the locality. I cannot see why this problem should not be overcome in a large area, by the intelligent allocation of catchment areas, and in a small area of course, there is no problem. Perhaps a plus and minus points system could be worked out on the basis of plus marks for low density, and minus marks for the opposite. As most of the country is committed to some form of comprehensive education, it is difficult to see why the powers that be seem bent on preventing its completion, at the same time sabotaging the potential of the existing schools. By definition, it is impossible to get the best results if the cream is skimmed off into another jug – in this case the grammar school. It should be all or nothing.

Among the children sitting at their school desks today are those who will in the future, be sitting at boardroom tables as leaders of industry, and even in government chairs from where they decide the future of the human race.

This thought poses another question. Is the man who has risen from the bottom, more authoritarian, more competent or a better chap to work for than, for example, one who has entered his job under the cloak of parental protection? One cannot say; it depends entirely on the individual. Some types promoted from the ranks make the worst kind of officers, perhaps dishing out to others what had formerly been meted out to them. On the other hand, the assumption that reformed poachers make the best gamekeepers, could be based on their having sampled conditions at the bottom of the pile, they become more considerate and enlightened higher up!

Perhaps this is an argument for a more flexible society. In spite of the millions of words written on management, direction of labour, etc., it comes down to few basic principles. Example is more meaningful and effective than exhortation or coercion. I think that the abuse of power by those in charge of affairs when capital exploited labour, reaped a bitter harvest when the positions became reversed. All work has a certain dignity, even shovelling cow muck, and workers should ideally be able to respect the job and the boss, and he should do the same in return. One cannot expect loyalty by dictation.

Human nature sometimes brings on despair, but hope (and faith and charity too, for that matter) is only just around the corner, so I don't take the pessimistic view of an acquaintance who thought that 'if you give a man a pair of boots he will kick you with them'.

I do keep on about wealth! Winning the pools, for instance, is what millions of us dream about for a few seconds every week. I don't do them now because buying the postal order was too much bother. I suppose I'm against gambling, and realise it's a mug's way of a diversion, but I don't mind a few bob on the Grand National, especially if my horse wins. Wealth is the magic factor that gives us freedom to behave in an uninhibited manner. To say what we like without worrying whom we offend, and to release us from jobs to which we are tied like a goat to a tree. Financial independence seems like an unattainable dream, even if the achievement is less enjoyable than the striving. *Is* travelling better than arriving?

It has always bothered me that in the Gospel according to St. Matthew, it states that it is easier for a camel to go through the eye of a needle than for a rich man to enter into the kingdom of God. Being a literal-minded chap, I note that this sentence does not qualify the rich man. It does not say that bad rich men cannot get to heaven or that good rich men have the passport. If we take the words at face value are we to understand that affluence is a corrupting, bad thing? If you feed a horse on too many oats it's possible that he may come out of the stable on two legs. Wealth, like power, is heady stuff for the members of our species who cannot carry their corn.

Wealthy people seem to worry much more about losing their brass than poor people worry about getting any. Once I was in a taxi, and having only the exact fare in my pocket I was unable to give the driver a tip. "I'm sorry I haven't any more cash for the tip", I said. "You want to get some mate", shouted the cabbie, "It ain't half useful!"

When affluence arrives, people seem bent on conserving everything – capital, privilege and even blood. My wife made the observation that it's mainly ordinary working folk that become regular blood donors. Well, some get more pleasure out of giving than receiving, and it's a good thing for the rest of us that they do!

Human nature is very queer. I like the story of the farmer who was discussing the relative attitudes of different animals compared to the human race. "Dogs", he said, "being servile, are definitely inferior. Cats, well, cats are superior; but pigs – pigs" he said, "is equal." None of us is immortal. It's only a glutton who would wish to live twice, and think of the conceit if that were possible. Some are arrogant enough on the existing trip. Just as some people take an obsessive interest in their ancestors (as if it matters whether you go back further than a couple of generations), others seem to be bothered about posterity, having a desire to be as well thought of by future, as by present generations.

Giving with thoughts of reward *must* be bad. If one is a genuine giver, any thanks that one receives should be a bonus and not a right. Andrew Carnegie gave all his money away, I'm sure for altruistic reasons. Quite unusually, he did not believe in founding a dynasty on inherited wealth – unlike the common practice of men in his position.

I read a story in the 'Dissent' article in the Observer recently, credited, I think, to a Mr. Seymour about a tramp who met a substantial landowner who ordered him out of the field. "Why?", asked the vagrant, "It's my land," replied the landowner. "How did you get it?" "From my ancestors." "How did they get it?" "They fought for it." "Well, I'll fight you for it then."

Returning for a moment to the subject of ancestry, here's another thing I've often pondered on. If we assume that the period of time between one generation and another is twenty five years, a baby born now would have, in thirty-six generations, a family tree like a huge inverted pyramid. The top line would contain 68,719,476,736 names and would go back to the time of the Norman conquest. As this figure is far in excess of the total population of the world at that time, notwithstanding the relatively static nature of the human race, it would seem that everybody is likely to be related to everybody else if one goes back far enough. Granted that many names in the sixty-eight thousand million would be written many times, it would appear that tracing one slender line back to William the Conqueror is largely immaterial. If one includes all the larks they got up to in the old days as well, it shows the futility of judging a person by who he is, rather than what he is!

So, if we accept the fact that money or the accumulation of same is a practice of doubtful virtue, the aforementioned cash should be used to experience and enjoy exciting things of beauty, culture and interest, and to further man's thirst for improving his intellect. Very high minded, I can hear you saying, but isn't that what the quality of life means? Unless we can latch on to some of it we might as well become a colony of ants, and who knows, we may be able to donate part of our brain to science even before we have finished with it!

For most people a new experience or interest in their lives comes like a new friend, by introduction. I personally, found over the last two years, great enjoyment in the work of two composers that to me were formerly names only. These two, Brückner and Mähler, were revealed only because my office at the farm is next to the coalhouse. After we had installed oil-fired central heating, and then found that the coalhouse was not being used, we converted it into a study/bedroom for our eldest son. He found it advantageous to play the works of these two composers during his swotting, and very loudly too, so my ear took a fair bashing, but I became richer for the hearing.

Another interest I have in a modest way is antiques, thanks to Marie, my spouse. In the early days I had no desire, time or energy, physical or mental, to absorb another subject, as I was working flat out, hoeing, ploughing and milking. When we were married we furnished the house with a mixture of secondhand stuff, and furniture my mother had discarded, hoping to buy brand new later on. Fortunately, before that stage arrived, Marie cottoned on to the fact that buying antiques with imagination gives one's possessions the chance of increasing in value – unlike new furniture which, with a few classic exceptions, can only depreciate. In addition, the old piece possesses an identity and warmth, having been part of the lives of so many people since leaving the workshop and hands of the original craftsman.

Thus it was that Marie started buying lowbrow antiques at auctions, junk shops and jumble sales, gradually relegating the real junk, either to the bonfire or to useful service on the farm as cupboards for tools, spanners, cattle medicines, etc. One chest of drawers made a super set of nest boxes for my Rhode Island Reds!

Looking back, one could weep at the thought of the objects slung out, as civilisation spread over the countryside before and just after the war. For example, all the brass bedsteads that I have seen blocking up gaps in orchard hedges, keeping the calves in, and the brass lamps that were thrown away when electrification reached the villages. I bought one twelve years ago for ten shillings, and it's now worth £30. Copper kettles and brass cooking pans, all in use in both country and town kitchens on the old triplex grates, now gleam with much polishing and have new functions as ornaments, or with the current craze for floral arrangements, either contain or act as accessories for a floral fantasy. Pewter measures, used in pubs before optics were available, now command high prices – so many were simply discarded. Chamber pots are passing over the Atlantic to the U.S.A. in a never-ending stream! The humble green glass paperweight now commands a great price alongside its much grander and more beautiful French counterparts from Clichy and Baccarat. Even old flat irons, with which housewives struggled only twenty years ago, are used now as bookends or doorstops and cost about 50p depending on size.

Every little girl had to produce a sample of her needlework in Victorian times. These mostly maple framed samplers are collected with great interest, and are often discovered in attics where they had been abandoned by the unwilling needlewoman, as soon as her mother no longer had control.

I have always liked pictures, particularly landscapes and those portraying animals, by such artists as Cooper, Stubbs and Herring.

I managed to buy, for twelve pounds, a nice portrait of a Breton girl, with a big gash across her face, damaged I think by someone gone beserk with a billiard cue. A friend did a smashing repair job, and it now looks quite smart hanging in our sitting room; people think it looks like my eldest daughter.

Another piece of luck occured when I was staying at the Park Hotel, Cardiff, for a radio programme. Coming down in the lift I accidentally pressed the wrong button, and instead of the ground floor, I emerged at the first, where there was a big room, arrayed with paintings, awaiting an auction later in the evening. As I was unable to stay because I had to attend the recording, I put a modest reserve on two lots – one a George Morland and the other signed by an unknown (to me) artist. Both depicted cattle. Later I learned that I had become the proud owner of both pictures. On my return to Ticknall, I cleaned up the Morland, but the 'GM' washed off immediately, obviously having been added later. The other, however, by the unknown Dutch painter, was valued at £300.

One of our first pieces was Charlie Barker's Windsor arm chair, with crinoline stretcher and made from yew wood, bought in a sale for £3 and now valued at £70.

If one has an eye for interesting bric a brac, it used to be possible to collect objects for shillings, as Marie did, using housekeeping money and sometimes what she could find in my trouser pockets.

Sometimes one can obtain a particularly prized possession, treasured for other reasons than purely financial. When Amelia, our second daughter was born, in 1962, we received a letter from a lady in the south of England, whose father was vicar of Ticknall in 1899. She was disposing of the contents of her home, and had in her possession an oak Dower Chest, with the inscription "1670 Ellen Moult", carved on the lid. Naturally we bought it, and subsequently christened our third daughter Ellen. When the six children are fighting over our bits and pieces after our demise, Ellen certainly has sole claim to one piece.

My interests in antiques and Fine Art had always been on the opposite side from the auctioneer, but in 1969 I had a unique chance to see how things were done from his end.

Derek Burrell-Davis from the Outside Broadcasts Department of B.B.C. Television Service asked me if I would be interested in acting as the chairman in a programme to be called *Going, Going, Gone*, to be recorded in August at Sotherby's Saleroom in New

Bond Street.

There were to be six programmes, all to be done at the Saleroom while it was closed for the month. Briefly, the game was a contest between two ladies and two men who were bidding for the various lots on offer. Each lot had a predetermined 'resale value' established by the Chairman of Sotherby's, Peter Wilson, who 'sold' the items from the rostrum during the game.

In addition, there was to be a novelty 'odd lot' of a more bizarre nature, like bits of granite from London Bridge. If the experts bought at above the 'resale price', which was the sort of price one would expect to pay in an antique shop, they would show a loss, and below it a profit. The last bidder of each lot would be invited to justify his bidding.

It was as complicated as it sounded. I like being chairman of things, but it is far more difficult being in that job than as a panel member in a game like *Call My Bluff*, which is like being on holiday by comparison.

In fact it was almost as nerve racking as the time Don Mosey asked me to take part in the televising of the *News of the World* darts championship at Bolton. During the interval, I was kidded to allow one of the darts maniacs to throw his arrows at a cigarette which I was smoking at the time, while my nose was trembling on the treble nineteen!

What's more being chairman involved giving the score, which is quite a tricky job on its own. I have given wrong marks on a number of occasions when conducting quizzes, and was involved indirectly in one memorable mix-up, when Katie Boyle, Jack de Manio and myself gave the title 'Miss Post Office Personality Girl' to the wrong telephonist! The reason I'm putting forward is that the 'electronic' scoreboard couldn't hear properly when the marks were announced ! !

Each week in *Going, Going, Gone*, two subjects were involved, and these included glass, antiques, English ceramics, Russian works or art, paintings, armour, firearms and many others. So every week I left the farm, and for a day became part of a dream world of Jacobite drinking glasses, figures of Budda, icons, Worcester dishes, fabulous objects by Fabergé, to mention just a few.

One of the opening shots was a close-up of a famous painting of the Stag Beetle by Albrecht Dürer, sold shortly afterwards to Alain Delon for about £68,000. I got a big kick from actually holding this small work of art in my own hands. Perhaps it's the nearest I'll ever get to such a valuable one.

I imagine the idea of the programme was to entertain those who might wish to be stimulated further into the world of *object d'art*. I can only assume that my function was to show that Mr. Everyman can derive pleasure from the subject, and to bring the discussion down to a common level if it became too esoteric. Unfortunately, the series finished just as I thought it got into a swing. Who knows what heights it might have reached if it had kept going and not gone!

The faculty of discerning and enjoying beauty and standard of excellence, is what may loosely be called taste, really another way of saying education and enlightenment of the five senses. Each overlap, but possibly each individual has them developed to a different degree, some senses being stronger than others.

This subject occupied my mind one day at a popular seaside resort, as I watched a jostling throng of all ages drift aimlessly from emporium to emporium, stacked with mass produced super-junk. Kitch, in fact. Nor is this exclusive to Britain. Useless items made for gazing at, seem to be available wherever mugs can be parted from their money. In all tourist spots one can find this kind of merchandise. Perhaps the only reason that the junk of our fore-bears seems not to fit into the category is because without the benefit of mass production, every item was individually made by craftsmen. Is this the reason why prizes on the bingo stall are now acceptable, and that appreciation and recognition of artistic objects is limited to the minority?

In former days, the Georgian period for example, which was possibly the most elegant of all, employment for a large proportion of the population – certainly girls and young women – was limited to gentleman's service. The 'taste' of their so called betters (actually, more fortunate employers through an accident of nature), was learned and appreciated by the employees in large households. For instance, a wedding at the 'Big House' would be conducted in great style, with the organisation of the proceedings following an established pattern, which ensured that all involved were mentioned in order of importance. Throughout the function, therefore, all the guests were at ease and perfectly aware of the order of things. In a more humble way, a wedding of one of the household staff would follow the same pattern. Nowadays, a gulf has appeared. Divorced from this inverted pyramid of influence, people have lost the rubbed-off effect of style. The gap has been filled by commercial interests, so that many people are uncertain and ill at ease in a formal situation.

Knowledge is a fine thing – I wish I had more of it, but it's fun finding out.

A farmer took his son to London for the day. They saw many sights, including St. Paul's Cathedral.

"What is that big church, father?" said the boy.

"I don't rightly know, boy" replied the Pa.

Then Buckingham Palace.

"What's that big house, Father?"

"I don't rightly know, boy."

On to Nelson's Column. Same question – same answer.

"Dad," said the son, on the way home, "I hope you don't mind me asking all these questions.

"That's all right boy, if you don't keep asking questions you'll never find anything out!

There is no end to the list of fascinating facts that I have discovered, just as I go round, stumbling across them so to speak. For example, do you know how North American Indians get their names. Well, every child is named after the most memorable thing seen by his parents on his birthday, which accounts for Sitting Bull, Rain in the Face, Crazy Horse, Laughing Water and so on. It's quite an idea for a quiz show, or it would be revealing to apply this principle to contemporary Britain. I wonder what names we would inflict on our defenceless babies!

Years ago, on an island in a warm sea, each inhabitant had an urn in his house. At the end of each day he reflected on whether he had been happy or otherwise. On the happy days he would throw a white pebble into the urn, and on sad days a black one. When the man died the quality of his life was judged by the proportion of black and white pebbles. Do you think this could be a valid exercise?

People who can live a full life are fortunate, and I have always been grateful that fate gave me the opportunity of seeing, doing and learning about so many things. I have often been asked what I like doing best of all. I can sum it up by the answer an old farmer gave, when asked what his hobbies were. "Thinking about what I'm going to do tomorrow", he replied.